Rae Taylor's book speaks to our identity, indeed to our very soul, which is deeply embedded in the lands we know and love. Her descriptions of the acute, heartbreaking loss of physical place and cultural space with the wounding of spirit that comes from that loss, are balanced by dynamic stories of hope and healing that provide nourishment and strength to continue to protect and live a wholeness with the lands. Her writing is firmly grounded in the land and waters of the American West, as she tells stories of people and community who provide stewardship, reflecting a reciprocity of being one with the land, creating what she calls "a homeland of spirit."

—Nancy C. Maryboy, PhD
President and executive director
Indigenous Education Institute

Paradox

My land is the torn and tender earth
harboring rocks and ravens as guide
punctuated with songs of wilderness
and pangs of destruction
full of breaking hearts and stalwart souls
laboring hands and interminable courage
fettered with travesties of ritual, insults of crime
stemming with unending faith and unheard of joy

Praise for THE LAND: OUR GIFT AND WILD HOPE . . .

"You can best serve civilization," Wendell Berry once wrote, "by being against what usually passes for it." By such a reckoning, Rae Marie Taylor is a truly civilized artist. Her passionate love of the Southwest and her grief at the destruction that has been inflicted on its landscapes and cultures shine through this book. It offers a deeply personal witness to the power of the land and the force of American ideals. In the midst of our devastation, she manages to celebrate. In the midst of sorrow, she finds a hard-won hope.

—Mark Abley
Author of *Spoken Here: Travels among Threatened Languages*

Rae Taylor's poignant and uneasy love affair with the American West testifies to the endurance of hope even as the narrow logic of economics, like a geologic force, erodes all things in its path.

—Bill deBuys
Author of *River of Traps* and *A Great Aridness*

One of the great attractions of the American Southwest is its stark blend of slow-moving natural beauty with frenetic human energy. Sometimes the two forces mesh well, as when ranchers and other land-based peoples work cooperatively with the land; but sometimes they clash, often with tragic consequences. Rae Marie Taylor captures the complicated relationship between people and land in the Southwest with a sweeping sense of history, humor, and insight. For anyone who wants to understand where this relationship is headed in the twenty-first century, I recommend reading this book.

—Courtney White
Author of *Revolution on the Range* and cofounder of the Quivira Coalition

Rae Taylor explores her love of the West in eloquent and gracious passages that weave together life experiences with wisdom gleaned from a lifetime of careful attention to the land. "A body needs the earth," Taylor writes, finding inspiration in Barry Lopez's Arctic Dreams. *"It is a question of habitat." For Taylor,* The Land: Our Gift and Wild Hope *is a habitat of the heart, carefully cultivated with tough love during years spent in Colorado and New Mexico. For the reader, the book is an invitation to contemplate your own relationship with the land and with the cultures that have formed you.*

—Page Lambert
Author of the best-selling memoir *In Search of Kinship* and
senior associate with the Children & Nature Network

Charmed, intrigued, educated, outraged, uplifted, enchanted, activated— all these soul states await readers who journey with Rae Marie Taylor through the pages of The Land: Our Gift and Wild Hope.

—Steven McFadden
Author of *The Call of the Land: An Agrarian Primer*
for the 21st Century* and *Farms of Tomorrow

Querencia, *a Spanish word that is hard to translate into English, identifies both the love of that particular spot in the world where one feels safe and grounded and the instinct to return to it.* The Land: Our Gift and Wild Hope *is a book about* querencia, *Rae Marie Taylor's return after many years to her beloved New Mexico only to find it, like so many sacred places in the American West, changed almost beyond recognition by development and materialism. But this is also a hopeful book, filled not only with an articulate call to heal the West but with the stories of people and organizations that are doing just that, healing not only the land but themselves. Taylor has a great heart, evident on every page.*

—Teresa Jordan, author of *Riding the White Horse Home*

May 2016

Michael,
*   Meet another of the Taylor sisters!*
Enjoy the tour of the West.

Love,
Harriet

THE LAND

Our Gift and Wild Hope

RAE MARIE TAYLOR

Bright
Shores
PRESS

QUÉBEC, QC, CANADA

Published by:
BRIGHT SHORES PRESS
874 Avenue Brown #2
Québec, QC
G1S 2Z4 Canada
http://raemarietaylor.com/theland

Book design: Angela Werneke
Cover and interior illustrations: Rae Marie Taylor

Text and illustrations copyright © 2012 by Rae Marie Taylor

First Edition

In excerpted form, parts of *The Land: Our Gift and Wild Hope* appeared in *Sustainable Santa Fe 2010* and in *The Return of the River* (Sunstone Press, 2010).

The Land: Our Gift and Wild Hope is factually accurate, except that occasional names and traits have been altered to protect privacy while preserving coherence.

Printed and bound in Canada by Friesens

PUBLISHER'S CATALOGING-IN-PUBLICATION DATA

Taylor, Rae Marie.
    The land : our gift and wild hope / Rae Marie Taylor. -- 1st ed.
    Québec, QC, Canada : Bright Shores Press, c2012.

    p. ; cm.
    ISBN: 978-0-9733962-2-5
    Includes bibliographical references.

    1. Sustainable living--West (U.S.) 2. Environmentalism.
    3. Traditional ecological knowledge--West (U.S.) 4. Culture--
Environmental aspects--West (U.S.) 5. Ranching--Environmental
aspects--West (U.S.) 6. Reclamation of land--West (U.S.)
7.Wilderness areas--Environmental aspects--West (U.S.)
8. Feminist geography--West (U.S.) I. Title.

GE198.A165 T39 2011   2011924650
363.7/0978--dc22   1109

1 3 5 7 9 10 8 6 4 2

For Kay

and our being born to the columbine,

the meadows, the horses and the Bowen,

our river, together

ACKNOWLEDGMENTS

The author is grateful to reprint from the following sources:

"Sabaaths 1998, IX," copyright © 2005 by Wendell Berry from *Given*; used by permission of Counterpoint. Excerpt from "An Ecotone, Not a Divide" by Julie Sullivan, published in *The Quivira Coalition Journal* (October 2006); used by permission of the author. Excerpt from "The Biological Effects of Noise on Wildlife" by Bernie Krause, published online in *Acoustic Ecology News/Issues*, copyright © 2001; used by permission of the author. Excerpts from "Sacred Places" by N. Scott Momaday, published in Sierra Club 1994 Calendar; used by permission of the author. Excerpt from *The Gift: Imagination and the Erotic Life of Property*, copyright © 1983 by Lewis Hyde; used by permission of the author. Excerpt from Hundertwasser, *Exhibit Catalogue of the Artist's Graphics* (Glarus/Switzerland), copyright © 1973; used by permission of Hundertwasser Non-Profit Foundation. Excerpt from "Ultimatum for Man" in *New and Selected Poems* by Peggy Pond Church (Ahsahta Press), copyright ©1976; used by permission of Kathleen D. Church. Excerpts from "The Ecstasy of Influence: A Plagiarism" by Jonathan Lethem, published in *Harper's* magazine, copyright © 2007; used by permission of Random House, Inc. "Witness" from *Rain in the Trees* by W. S. Merwin, copyright @ 1988; used by permission of Alfred A. Knopf, a division of Random House, Inc.

Grateful acknowledgment also is made to Eleanor and Phil Bové, Tuda Libby Crews, and Dominique Mazeaud for permission to include excerpts from the author's interviews with them, and to Estevan Arellano for permission to quote from his lecture at the Quivira Coalition conference in 2006. Warm thanks as well to friends who have allowed their statements to appear.

By way of thanks

Heartfelt appreciation goes to many people who have sustained my will for writing this book.

First to my editors: Mark Abley in Montreal for his respect and understanding of my intentions and his finely focused editorial comments; and A. Kyce Bello in Santa Fe for her rich exchange and stimulating editorial response. Much appreciation goes to Margret Carde for her astute and valuable suggestions.

My readers, Thomas Antil, Brad Holian, Barbara Stanislawski, and Anita West, were repeatedly willing to take a fresh look at new chapters. I am grateful for their eye for accuracy and the hours of thoughtful reading and motivating exchange on our land, its dilemmas, and even the cosmos that have so enriched the process. I'm equally grateful to Eleanor and Phil Bové, Tuda Libby Crews, and Dominique Mazeaud for their readings and sharing of their time and knowledge in interviews, which they made a pure pleasure.

Special thanks go to Ernie Atencio for his support and for having drawn me to the people of the Quivira Coalition; and to Reina Kohlymeyer and the Buffalo and Corn Dances of Jemez Pueblo that comfort and renew my spirit both on my frequent visits and even when I am far away.

I bear a debt of gratitude to N. Scott Momaday, Lewis Hyde, Wendell Berry, William Kittredge, and the other authors the reader will find quoted here, whose writings have sustained my thinking and hope for years.

Additional warm thanks go to Brad and Kathy Holian for the gift of many weeks in their home during the writing, and for sharing their horses along with their deep concern and knowledge about our homeland's dilemmas.

Conversations with people I spoke with intermittently about the book often helped clarify my intentions in the way that only real dialogue can. I would like to extend my appreciation to my National Park friends as well as Anne Beer, Catherine Chase, Peg Crumbacher, Paulette Frankl, Harold Garde, Dana Hearne, Patricia Jean Manion, Jim Mafchir, Fran Nicholson, Lewis Poteet, Jutta and Bertrand Valois, Bob Vukanovich, and Courtney White.

A bow to Angela Werneke, designer, for her fine technical assistance and her thoughtful artistic insight.

And a long bow to Ellen Kleiner of Blessingway Authors' Services in appreciation for her creative insight, her dedication, and for bringing not only her extensive publishing experience to the job but also her marvelous capacity to stay caring and positive through the trials of publication.

Finally, my heartfelt gratitude to my sisters, cousins, and community of friends in both the American Southwest and Quebec who have offered invaluable encouragement, insight, hospitality, laughter, and comfort during the years of writing. You keep my life whole by always welcoming the nomad I have become.

Contents

Introduction

In *The Land: Our Gift and Wild Hope*, Rae Marie Taylor remembers with aching beauty her experience of the western landscapes of Northern New Mexico and Colorado as an energetic young girl roaming freely through a wild and abundant land. Here memories are inextricably tied to family and community living with, not on, the land.

Her book is a sequence of individual essays that each stand alone yet together form an extended single essay. They recount the complex pull between homelands: the author's expatriate home in Quebec, and the American Southwest, her home since childhood. An artist and writer, Rae Marie Taylor has taught language, literature, and creative writing in her primary residence, Quebec. Until recently, she and her family were blessed to be able to return to their history with a solitary cabin in the Colorado mountains. Making return pilgrimages to New Mexico to write, paint, and sculpt, she still joins a circle of community around the Pueblo dances, and talks into the night with old friends. As she says, "My life has been rhythmed and punctuated with return visits and extended stays in the wilderness of the Rockies and the high desert in New Mexico."

On one such return, when Rae was in Santa Fe's Cerro Gordo Park for sunrise, she was moved by ravens appearing out of the morning glow. The author marks the moment as the epiphany that birthed this book. "I saw that the privilege to have experienced a long and safe relationship with the wilderness in a culture that valued it, brought with it the responsibility to witness the beauty, the wonder, the health of the land I loved, and—Lest we forgot!—the necessity of our relationship with it for our humanness." She knew that she must tell her truth about what was happening to her beloved West, pointing to W. S. Merwin's poem, "Witness," as a warning that the tale must be told "before it is too late."

I want to tell what the forests
were like

I will have to speak
in a forgotten language.-

The Land: Our Gift and Wild Hope recounts Rae's repatriation to New Mexico and the pain engendered by the reality that her childhood home had changed; the land of her memory still exists but only in fragments.

This work is a journey of the heart to find a way back "home," motivated by the mystical experience of the family cabin and the New Mexico of her early memories. The road is complicated and her return is tenuous because she cannot earn a living sufficient to support even a small existence in this Santa Fe of rampant real estate development and high prices. Her physical poverty is demeaning and, like many native New Mexicans, she is forced to live as a foreigner in her own homeland. At each turn, the need for new understanding is personally challenging.

Throughout this journey we learn that being with the land, its flora, fauna, rocks, rivers, and wildness, is essential for establishing a personal relationship with the being we call Earth. We discover that people who are moved by the land carry it inside themselves, within their cellular structure. We learn that the wild's external vistas resonate with our internal topography, awakening the spirit, which responds with awe and gratitude. Similarly, openness is critical to western hospitality. One key experience in the book illustrates how a new neighbor's refusal to acknowledge the author's wave of hello can literally stop the flow of generosity between neighbors. Belonging requires openness: individuals opening in generosity toward each other; cultures respecting one another; ritual reverently repeated with and for the land; and a science of empathy that fosters land-sustaining practices.

For Rae Marie Taylor, the heart of the problem lies in the isolation of individuals or cultures from the direct experience of and reverence for the land. This isolation is created by greed for wealth, fueled by the need for creature comforts and the security of the known. Anything unknown becomes "other," and this "other" must be excluded, whether it be people, animals, or wildness. Neighbors, wild animals, and unfamiliar cultural practices become irrelevant. New gated communities and drivers who impatiently honk at slow-moving traffic tear holes in the Old West fabric of connection among people and with the land. Separation erodes communities. The land suffers and, with it, the people suffer.

The Land: Our Gift and Wild Hope is a journey told in a woman's voice. The language is intensely personal. The author openly searches for the fragments that will lead her back to the feeling of being alive—physically and spiritually—in the silence and wonder of the wilderness. At times speaking through the perceptions and voice of herself as a young child, sometimes taking the probing voice of a reporter interviewing artists or ranchers, sometimes speaking from her inner voice of wisdom and experience, Rae Marie Taylor adapts her voice in a way that most directly conveys the heart of their message. She pieces together people, places, conversations, memories, questions. She highlights key concepts, quoting authors who also share insights they have gleaned from pain caused by loss of relationship to the land. She carefully mends the seams and knits together the edges, filling the empty places with her own poems.

Following the thread of language across time and into her own psyche, Rae Marie Taylor weaves her own history, sharing with us a story common to many in our culture. With each tug of the thread, with each new insight, we discover that generosity must apply within oneself, between ourselves and our memories, friends, communities, experiences, and with the land, even as we resist new and raw intrusions that tear at the fabric of our own remembered truth. In the end, we look back at a patchwork of form, a quilt whose patterns reveal a deep theme: the search for reciprocity between a timeless land and the people who cherish it.

—Margret Carde
Artist, writer, attorney, and nuclear safety advocate

I. MIGRATION

MIGRATION
Ceramic, stone, and found objects

ONE

E Pluribus Unum

The night Russia invaded Hungary I sat comfortable, attentive, in our Denver home, next to my mom on the green and white couch, as she and Dad read the newspaper and spoke. I understood that invasion meant war. Unjustified. How could this be? Could it happen here in Colorado? My dad explained we lived in a country of tolerance. To the ten-year-old that I was, the United States had always seemed a vague, distant place with no mountains, embodied by faraway cities with names like "Philadelphia," "Washington, DC," and "Boston," where my dad had gone to school.

On that historic night in 1956, I learned that my dad believed the United States was a democracy. As such, we would not invade unnecessarily. We respected differences because our country was founded on an ideal called *e pluribus unum*, "one from many," one country from many lands, he thought, one people from many peoples. This saying was inscribed along with the powerful eagle on our country's crest, the image of the protector eagle borrowed from the Mohawk. Reassured, I believed. For the first time, I embraced the notion of a country larger than the high-altitude plateau of dry plain and mountains thousands

of feet high that I already knew. E pluribus unum, I repeated to myself, one from many, the motto settling into the core of my love for my
country.

The next summer, our parents piled all seven of their daughters into
the green Woody station wagon and, instead of heading for the mountains as we usually did, drove us south to New Mexico. An even drier
land than Colorado, New Mexico was cherished in my mother's history
for being the home of Gen McBride, her valiant, independent college
roommate, and for being the the place for us to see Santa Fe's La Fonda
Hotel, where my parents honeymooned in 1939.

After songs and games on the road, and the happiness of the long
drive through canyons and along the Río Grande, our family trundled
into our little casita at La Posada, a hotel in Santa Fe, the old New
Mexican capital. Quickly my twin Kay and I discovered the pool where
our four teenage sisters (how I admired them!) strutted in their new
bathing suits, so "cool" in a 1950s way.

It had been earlier in Taos, though, where we had been keen to find
the plaza at the heart of town, so small, its dusty buildings seemingly
comfortable and very old. Several Taos Pueblo men were relaxing there,
their blankets beautiful in the sun, the kind delight in their eyes welcoming us children. The greetings were equally eager and warm at El
Mercado, the general store on the plaza, run by the McCarthys, friends
of the family, who had been in business since their grandfather had set
up the first trading post on the plaza decades before Taos became an
American town in 1934. In the store there was the fascination with hardware, the texture of fabrics, bright ribbons (ribbons!), and candy. But
more than fun and delight would reveal itself in this state affectionately
called "The Land of Enchantment."

Built of clay and vigas, Taos Pueblo sits on the plain below the Sangre de Cristo Mountain range, as it has for over a thousand years. The strength of the mountain peaks shimmers in the high altitude above it, sheltering Blue Lake, this Pueblo's most sacred site. On the day of our family visit, though, I hardly noticed the Rio Pueblo cascading down from the lake, joining, as Native people would see it, the two large plazas of the multistoried village, alive and somehow familiar. That day it was a man and his story that drew my attention. My big sister, Peg, and I had stopped to see the crafts two older men were selling. Jewelry? Pottery? Trinkets? Traditional drums? I don't recall. What mattered was the story one man proudly told us of his father fighting Kit Carson during the Indian Wars. He was animated. I listened, engrossed.

When we left the Pueblo, I mentioned his story to redheaded Peg, who dismissed it "because," she said, "he'd been drinking." Not as sophisticated in the ways of alcohol as she, a teenager, I stayed quiet. I believed the man. Fiercely. I knew the story he had given me was true. Back then, wars seemed so far away to that ten-year-old. Both World War II, when my dad was a naval officer in Europe, and the wars against the Native peoples fought on our own country's ground that this Pueblo man knew so intimately, seemed like the affairs of old men. My dad, who told few stories of his war, and this talkative son of America's war were alive. Well, I thought. Safe. E pluribus unum was intact.

In this land far from the United States, the air itself, the wide expanse of sky and dusty earth, the warmth of the adobe villages themselves—all breathed with the knowledge of the alert peace of the wild I under-stood. People we didn't know seemed to walk with the same knowing. I had thought this was limited to our mountains, our relatives, our friends in Denver. Now my country had expanded to a more encompassing reality, a larger, older world of Pueblo, Spanish, and Anglo villages and customs, built whole with the clay of the desert.

The family rode north, back to Colorado again, the landscape accompanied by the sound of our singing "America the Beautiful." My America. We reveled in thoughts of our mountains' purple hues, our brilliant skies, the fertile and desert plains. Cocooned in the car I felt a new belonging, a new comfort that this enlivening world we had just encountered knew, as we did, the sacred in the towering peaks, the breathless silence of the changing light, the gift of high-altitude rain.

Curious now about the world, growing up only whet my appetite for even more distant lands when a new language set a new direction. The impetus came in high school when juniors were required to study a modern language. A teenager wanting to be different from my twin and our friends, I chose French rather than Spanish. The teacher was negligent, too old and tired to care, and certainly not the handsome personality of the Spanish teacher, but I found myself thoroughly enjoying the sounds and sentences of the language. The decision to follow through with French studies in college in Denver came easily. Here there was another requirement: to spend at least one semester in a French-speaking culture. My hunger for the world ripe now, a cross-country car trip heading north with a fellow student brought me to Quebec, Canada, where I was quickly enthralled by the welcome of the Quebecois, the St. Lawrence, their mighty river (so much water!), and the wealth of another North American history inscribed in the city's seventeenth-century stones.

I embraced this culture, a new geography, and a hospitable people who, like westerners, seemed other, different from their larger cultural landscape, this time of Canada. A place of bright snow, maple sugar, and ample forests, where rivers freeze, the débacle of ice floes thunders on the shores of the St. Lawrence in the spring, and the tree's sap is still for many months of the year. *Nunustiaq*, "the beautiful land," is what the Inuit call their cold expanse of land above the 60th parallel. Quebec City, where I migrated in 1968, is a bit further south, but still on the cusp of the subarctic. Harboring a predominantly rural society until the 1960s, the St. Lawrence and the Laurentian Mountains nearby impose extreme winter conditions, blue and white, a different kind of desert, I have often thought. Both worlds, my Southwest desert and the artic North exposed to, surviving with, the immediate demands of the earth and the impositions of climate. In this life, I speak French and join in a productive, even protective, relationship with a more urban society, enjoying the joie de vivre of an energetic loyal community of friends, comfortable, above the poverty line.

I had migrated north for the love of the French language and found I loved the vigor of the people. Both mesmerized and energized by the challenges of the cold, the quantities of snow, I became part of the fabric of their society, participating in its Quiet Revolution of the late 1960s and 1970s when, with colleagues of my generation, we established the CEGEPs, a democratic system of community colleges to replace the elitist postsecondary education system that had reigned until then. Over time, I might weary and suffer from cabin fever during long bouts of -20 degrees Fahrenheit, but the bonding of people in pride over the rigors and beauty of this cold expanse redeems those moments and carries a familiarity. It is akin to the land knowledge people share in the Southwest when faced with the challenge of drought and fire, as well as when we revel in the vast, dry, spacious beauty of the desert plateau or the majesty of the Rockies. Beautiful lands.

One, though, has a greater pull. It is where my breath and cells mingle with the wind and spruce sap of home. There the word *land* itself resonates with awareness of the waters, watersheds, and beavers; trailheads and horses; dirt roads and paved, a rare mountain lion sprinting across. Even teaching thousands of miles away, the knowledge of its spaces sustained me for years. Was it my parents' aging or my intuition that much was being lost in the wild there, the sense of the West's strength fading with my own, that prompted my unbridled need for return? I wasn't sure.

I had been active in the College's teachers' union, in an environment where the educational community collaborated productively together. In the modern languages department my colleagues and I founded, we developed new pedagogies, created new materials, and founded a language center, a space for our students' freer access to language practice, materials, and support. Eventually, though, the collegial '70s were over. The community strained and stressed under the pressures of power politics. Deeply worn, the collective will was torn apart in the transition. In the thick of the fray as head of the department at the time, I attempted to protect our language teachers and students from the worst impacts of the institutional conflict. Over a one-year period the incidence of teacher illnesses increased by 40 percent, the head of Human Resources committed suicide, my own health suffered, and I dealt with a bout with cancer. Accompanied by the vital impetus to paint and write poetry, the successful treatments to restore my health were followed by the need to restore in situ the relationship with my home ground. After twenty years of snow and teaching, the Southwest drew me back. My mother would

often say, "The mountains are calling," those boulders and rocky peaks, trails and streams, the nudging deer and the plunging, playful raven. Spending more and more time with them on vacations home, I eventually responded with an exploratory trip, spurred on by a deep recognition that came through my art.

In the summer of 1985, I was living in the Quebec countryside when an unusual painting was born, sudden and intact, of a feverish concentration in the studio. As the image dried, I recognized Mesa Verde, home of the Pueblo peoples' ancestors, in the browns and blues of its abstract form. A few months later it was followed by an equally urgent state that produced a series of fluid, dancing figures new to my style. At the time, I had not yet visited Mesa Verde National Park, although the year before, when I had been spending summer vacation restoring myself by homesteading, painting, and writing at our family cabin in Colorado, a childhood friend suggested I see its ruins. She also thought Georgia O'Keeffe would interest me. Until then, having been more influenced in my artist life by European traditions and abstract expressionism, I had paid only vague attention to this displaced New York artist.

The following spring I was perishing in the gray days of a northern April, winter hanging on as usual, dirty snow not melting fast enough. Ah, April, that cruel month! Nothing seemed to revive my spirit. I knew I needed to respond to this state that was as spiritual as it was physical, but what was it I needed? During this period, a sketch of a southwestern-looking courtyard came of its own from my hand. Georgia O'Keeffe had died a month before; only much later did I realize that the drawing was of Georgia O'Keeffe's courtyard, now famous, but at the time a walled place I had never seen. New Mexico was calling.

Over the previous year I had been languishing for home, but unsure about where home was. Was it, perhaps, where one of my six sisters lived? I had had a wonderful exploratory visit with my younger sister, Anne, in California at Christmas, but it was clear that California was not quite the place, nor affordable. Certainly it was not Denver, which had become too big, too urban to accommodate my desire for the expanse and wild of home. The courtyard drawing was enough to remind me of the flush of the earlier painting and my friend's admonishing me to visit Mesa Verde; enough, too, to remind me of my childhood love of Taos. I set an enlivening goal, reviving my energies. A trip that June, of 1986, would be an exploration, with Taos and Mesa Verde at the heart of it. With the help of stand-by fares, so cheap in those days, I flew south toward the Rockies.

As luck would have it, my niece, Christine, cheerful, thoughtful, and an experienced camper, was available to accompany me on the journey. In a family car we headed south on I-25, stopping halfway between Denver and Taos to camp overnight at the Huerfano campground. I was grateful for the ease with which we pitched tent together, with no unnecessary talk, in the mutual understanding of the peace around us. Waking refreshed the next day, we rose and packed up, quietly entering the silence of the morning, the perfection of the mountain air. Accompanied by the shafting light on the mountains as we drove, we entered my land, the spiritual territory I had recognized in the painting. I had crossed the boundary into home. Our road took us on to a recognizable peace in Taos and to refreshed amazement as we lingered on the bridge over the dramatic beauty of the Río Grande Gorge. A walk through the pink desert farther on, at Ghost Ranch near Abiquiu, led us to a quiet moment at Georgia O'Keeffe's courtyard. From there the pilgrimage circled up again to the southwest corner of Colorado.

Mesa Verde National Park is a major archaeological resource and, especially, an ancient place where one is easily awed by the sense of the ancestral Pueblo presence in their finely constructed dwellings built a thousand years ago among the alcoves and overhangs of the gold and red cliffs. Like many visitors to these villages, my niece and I followed their smooth hand-and-toe trails worn over centuries into the sandstone. Hiking and climbing are always invigorating, and even more so among the ruins whose spirit remains alive. The rangers too were welcoming and generous with their knowledge as they stood greeting the visitors amid the soft stones. Recognizing our enthusiasm, Ernie, a calm black-eyed anthropologist among them, quickly became a friend and warmly suggested we hike the petroglyph trail.

The trail is bordered by a canyon wall with a long, animated row of migrating figures pecked into the rock. Exposed to these stimulating drawings—dramatic graphics alive on the cliff face—I experienced an immediate and moving continuity with them and their story of migration. The Hopi speak of these figures, telling of one clan who had left until the Whipping Kachina brought them back. The rock art figures spoke on their own, had presence still after a thousand years. I identified. These markings that record human survival and story rooted me in my land again, renewing my sense of belonging through, and in spite of, my own migrations. These petroglyphs also gave me a tradition as an artist. That human tradition sprung from the fundamental need to witness, to inscribe one's story. And a new clarity about the necessity for a human culture to know, manifest, and carry on its symbolic meaning.

Practically speaking, though, what would this encounter mean for moving home now? Spiritually, yes, I had come home on the petroglyph trail. But one can't live at Mesa Verde. Or could one? With such musings, Christine and I continued to Chaco Canyon to experience the labyrinth of rooms and large kivas in the village the "old people" had built there. Much to our surprise, we met Ernie in the campground. Away from the crowds, I was touched by his lean, handsome strength and knowledge of the canyon's people. We moved together among the ruins in wonder at our meeting and our deeply shared sense of the life of this land. After camping with Christine and me a little while longer, Ernie's road diverged back to Mesa Verde, ours to Denver. Taking leave of each other for the time being, this good man and I were willingly on our way to being a couple. The Southwest had been generous beyond my expectations. Flying back to Quebec, I was more than relieved, knowing again where home was.

Yes, it turned out, I could actually live at Mesa Verde for a while. In his letters over the year, Ernie signaled the way. The following summer, I settled into a season at Mesa Verde, working as a volunteer for the park's archaeology research lab and bookstore. My relationship with Ernie had deepened, in joy and hope for the future, and we embraced the opportunity to be together in this place we both loved. Our homes were just what I could have hoped for: first a hogan, a traditional Navajo dwelling, in the Mesa Verde Navajo workers' circle; and later a small "American" house in the rangers' neighborhood. "Pursuing a path of joy," as Ernie later described it, I lived, my spirit in rhythm with the beauty of the southwestern mesas. While my ranger friends educated tourists in the ruins open to the public, I was fortunate to scale cliffs with a team of archaeologists. The mornings were fresh, the team members welcoming and focused as we made our way to various sites, most often on park or hand-and-toe trails or, once, up a twenty-foot ladder we had carried in, admiring the yucca and scarlet penstemon blooming along the way. My first—and only—experience rappelling brought us down from a tiny site tucked under a cliff, and on one delightful occasion I was called to be flown in by helicopter to reach a nearly inaccessible site where archaeologists were debating whether the markings in a small cave were actually pictographs. Originally, perhaps high enough for the shorter ancestors to stand in, now the cave was just big enough to sit up straight in, on the layers of a large, round, and, fortunately, long-abandoned rat's nest. My drawings would show consistent or inconsistent patterns in the markings, which would help the archaeologist determine whether these chalk-like triangular shapes around the wall were rock art or bird droppings!

Immersed in ancestral Pueblo dwellings, privileged to be doing archaeological illustration of petroglyphs in the backcountry sites for the lab's plaster survey, I was at peace climbing, drawing, wandering in the wide, bright expanse of this land so rich with kiva and stone and people who knew and cherished it. Shared knowledge and attention to the importance of the lively rock art connected not only Ernie and me but a whole community to those ancestors carrying our human DNA who had walked, gathered, ground corn, irrigated, and hunted in this land a thousand years ago.

Although deeply nourished in my need to reroot in my western land, I realized that "the mesa," as it is affectionately called, could only be a temporary home. In spite of the beauty and depth of all we shared, I had to admit to the fact and the shock that Ernie and I weren't able to maintain a healthy relationship. At the end of the season, I headed for family in

Denver, staying a while to rethink and regroup. It was clear that whether or not I wanted to leave, I would have to go back to Quebec to earn some income, then return again. But first, before driving cross-country toward the northern winter, I needed to make a reconnaissance trip to Santa Fe and Taos, not suspecting what a moving experience it would be traveling this time with my niece, Susanne, and cousin, Katie.

For Katie   For Susanne

Tender Friends

> *are you my daughters, my nieces?*
> *surely you are my graces and my clan*

silent understanding of coral skies and owl's flight
widens my world, adds breath to beauty
threads us through with sky and earth,
wonder sustaining us
whole

gentle voices attentive to my heart
weave new beads of awakening light through the fabric
of my parting sorrow
I am touched by the cloak of your protection,
unexpected
> *lace*

Tender friends,

> *are you my daughters, my nieces?*
> *surely you are my graces and my clan.*

Just outside of Santa Fe, at Katie's request we visited Mildred West's small farm. An old friend of my cousin's, Mildred, eighty years young then, welcomed us warmly and encouraged us to spend the night. In the morning, I meandered down the hall toward the kitchen, replete with the peace of a night cozily tucked in under the black and starry sky. Already awake and at the table, Grandmother, as most who knew Mildred called her, sat quietly dignified with her long white braids. Sinking down onto a chair by the warmth of the woodstove, I confessed, "I never want to leave." Her voice was kind and matter-of-fact when she spoke, saying, "You can stay here when you return until you find your own place."

This was unexpected yet felt so natural. I accepted. The easy western hospitality that I'd known from my childhood was being extended to me. So I saw that Santa Fe, rather than Taos, would be my new home base in the next stage of my return. A wiser occurrence, I felt, knowing by now that if work was scarce in the Southwest, a region never part of a North American industrialized economy or infrastructure, the small capital city offered more opportunity for survival income than the even more rural Taos. Similar to Canada's Newfoundland, where a cash economy was late in arriving, the beauty of Northern New Mexico was that it had never been invaded by industrialization and had not yet seen urbanization (apart from the limited boundaries of the world of Los Alamos, where the atom bomb was invented). Such rural freedom, though, to live in the immense beauty of a landscape so many considered sacred, had its downside: mainly, no jobs or so few, and poorly paid or not at all.

The winter of 1988 was spent back in Quebec, teaching at the college, gathering funds. Renting a room in my old neighborhood, my three pieces of furniture in storage, I thoroughly accepted this transitory state, passionately, unquestioningly committed to returning to New Mexico in June. Over the semester, my Quebecois friends and I spent time as usual in the lively willingness to celebrate each other's lives and children; gathered again for the pleasures of hiking and skiing in the magnificent pine and maple forest on the rugged, rocky trails of the Laurentians. But I missed the daily presence of tall peaks at the far end of the street, the yucca and sagebrush, even the bothersome tumbleweed. I longed for my own knowledge of the lay of land and my ease in it, the crisp, clean air, and the Native presence clearly visible. I knew now I was too far from the sustaining rhythms of the Corn and Buffalo dances and the comfort of family. Skiing along the cliff above the St. Lawrence, noticing that perishing homesick feeling, the simple truth came clear: I so love my Quebec friends, this north's human warmth. I have needed them to respect and know myself. I've embraced Quebec's brilliant cliffs and its tides, but they don't sustain my cells in the same way as the high, dry trails of home. That space, that tall, mountainous, and wide plateau, the common knowledge of the soft, dry earth, my geography of shared reverence.

The students kept me busy over the semester until the day came for my third cross-country drive in a year and a half. Not a solitary exploit this time. Long curious to see the arid land of the Southwest, my scientist friend, Geoffrey, would accompany me. It was June. I had equipped myself with a wealth of determination and my savings of a few thousand dollars. My confidence came naturally. Having succeeded in making a good life for myself as an immigrant teacher in Quebec and having managed well in spite of student poverty in France, another foreign country, while writing my master's thesis, I did not doubt. I was a working girl who had made her way in a foreign land. This time I was going home. Surely, with the tradition there of housesitting, coupled with my willingness to work hard in a "day job," I would find the means to stay, produce my art, and eventually have an apartment or even a house of my own.

Thom, my only American friend in Quebec, came to breakfast early on departure day equipped with a banner reading, "Santa Fe or Bust." Laughter, hugs, high expectations, and warm goodbyes circled around as we fixed the banner on my beige Mazda GLC, a good and practical hatchback. It was during an intense heat wave, and in lieu of air conditioning, Geoff and I hung wet towels in the windows and refrained from too much talking as we drove several days across the country, heading for Colorado, "where the West begins." The road led past the sunflower fields of Iowa, down through the sage plains of Nebraska, to an enthusiastic reunion with cousins in Fort Collins, Colorado. A short visit with my parents at the old house in Denver followed. Tense. If my mother understood the necessity of my journey, my father, already mistrusting of the artist's life, was much less than enthusiastic about my leaving a good teaching position to head for an unpredictable future in what some people disparagingly called the land of mañana.

The Mazda headed south again on Route 285. One more day on the road, taking in the sky, now fresh and dry, the pronghorn antelope roaming on the plain, the vivid peaks (how proud I was to point out the many "14ers" among them!). Before we stopped to bask for a couple of hours in a hot springs along the way, Geoffrey remarked I was suddenly invigorated, my eyes brightened, freed, my face softened, my gestures more expansive. I was not surprised. Besides the yucca standing stalwart, there is something restorative about this land, sweeping and long inhabited.

We were west of Huerfano campground, reaching that invisible northern boundary of the ancestral Pueblos that was also the border of Mexico until 1821. This was home territory. And I knew where I was going. Over La Veta pass, down to Taos and a sunset stop on the Río Grande Gorge, on through Taos Canyon toward Santa Fe, with the Río Grande and my childhood memories of it streaming by. Greeted by the dogs and the brightness of the summer, we pulled into Mildred's farm the next day. I was relieved, jubilant, and confident.

The farm is nestled in the prairie south of Santa Fe, where, until millions of cattle grazed free in the late nineteenth century, wild grasses grew tall. I indulged in the daily joys of roaming among the blue grama grass (emblematic of my West) still thriving. Hiking down the arroyo (a dry riverbed), I heard echoes of La Llorona, the grieving ghost's cries. Meals were a chance to listen to Mildred's stories of teaching at Shiprock on the Navajo reservation later in life, or of her arrival as a young woman from Ohio, crossing the country in a Model T Ford with her husband Hal and, as she liked to point out, a pair of red shoes, in the 1930s. A true homestead, the door at Mildred's was always opening to visits from her large clan: sons and their wives and ex-wives, their children, and Anita, a granddaughter from Oregon with a great-grandchild on the way.

Rural chores, feeding the chickens and dogs, commiserating over the owls' diet of rabbits or, unfortunately, cats, and quiet talks of the phoebes' brood on the porch were punctuated by my own excursions back and forth to town in search of housing and work. Late summer days were absorbed into the rhythms of the clan's visiting, laughing, and eating that good green chile stew, piling wood for the winter, teasing, generally helping, all of us gathering around Mildred, the feisty matriarch who both complained about and enjoyed the activity.

Moving into town at the end of the summer, good fortune seemed to smile. A small grant would allow me to rent a large studio with a tiny living space for several months. My first invitation to exhibit my artwork came for my paintings that had integrated petroglyphs. Discovering work by Emma Whitehorse and other New Mexican artists at the capital's Governor's Gallery show, I was inspired and grateful to be welcomed among the artists who connected with the land. One morning at my first job, my boss kindly offered me a dog, knowing my kitten had been killed by two semiferal dogs next door. "What's his name?" I asked. "Cisco!" she said. Cisco, a blonde hybrid, was born wild in the mountains near Bandelier National Monument and would be a bright, somewhat unruly, light on the trails with me from then on. In line with my work at Mesa Verde, an archaeologist friend facilitated my doing lithic identification in the Southwest Regional Parks' archaeology lab. I began as a volunteer, the privileged way of entering the workforce. Not until Christmas, when possibilities of being paid for the meaningful work died out, did I realize how fragile my economic straits could become.

Looking back, I see myself juggling, as most working people did, several five- to seven-dollar-an-hour jobs, eager to stay, eager to keep a roof over my head, eager to become part of the community. I worked enthusiastically, serving cappuccino, leading archaeological tours, doing data entry in offices, teaching in my studio, managing literary conferences, or selling books and Christmas decorations by Pueblo potters in local stores, all to stay afloat on or below the poverty line. This was typical of what is called the "Santa Fe shuffle," the dance of multiple jobs, most often short lived, usually part-time, due to a seasonal business climate and an economy closer to the Third World than that of the states to the north, west, and east. I had, naively perhaps, looked forward to a simple life working at one single, adequately paid, part-time job while being productive in my art. With her trusty sense of humor, my eldest sister Mary, who had known poverty herself from years in Egypt, encouraged me to stay "lean and mean!" I was stymied to learn that only two of the two hundred galleries at the time accepted work from artists living in the city. Commiserating over and joking about our necessary plights, seasoned survivors would remind me that Mabel Dodge Lujan and Mary Cabot Wheelwright—admirable American women in New Mexico history—had been independently wealthy. Among the women artists we knew, it did take an independent fortune, a trust fund, a good divorce settlement, or a traditional marriage with a partner willing to offer room and board for a woman to steadily produce her art here as I had hoped to do. Without these and with dwindling savings, I still wanted to stay. If the income was sparse by my former middle-class standards, much of the work was all of a piece with the meaning of the West, linking me with people who not only cared but were also knowledgeable about it.

"The shuffle," however, was not really the full extent of the threat. Soon I was stricken with a reality beyond Mildred's farm and my own survival challenges: the realization of the loss of land and custom to a new colonization by late-twentieth-century greed. As I reentered this homeland, I was often seized with fright that both the settled Europeans' and the Native peoples' indigenous insights, finely tuned to the complexities and fragility of my desert/mountain homeland, were being lost. Mildred's trees, the moonlit peace, and her beloved pond were still available to me. My fun-loving, redheaded sister Peg now lived in Santa Fe with her husband Paul and their children. We delighted in taking spontaneous drives and hikes together when "the mountains were calling." At the ceremonial dances in the pueblos, I could gather with newfound friends, but in town, in spite of the dignity I tried to maintain day-to-day, something was missing. It was the recognizable fabric of my western culture. The home that had been rooted in the shared relationship with the land was being reduced to images, houses, and products for consumption by the extremely wealthy, a new version of "Santa Fe style."

This term was first coined in 1957 when the architect John Gaw Meem, once a newcomer himself but respectful of the land he had come to, was successful in obtaining protection for the city's traditional Pueblo Revival and Territorial architecture. In the hands of the 1980s migrants, however, the sense of protection the phrase carried was lost. Not only were adobe dwellings taken over. Cowboy hats that meant competent running of cattle or skillful horsemanship, the clay pots that evoked the sacred carrying of water and ritual gestures with the corn, the jewelry that was synonymous with pride in belonging to the land or interacting with its bear or deer, all had become sophisticated trappings of the new migrants, expensive and illusory objects rather than effective symbols. Their power to bring us together was waning. When only the image counts, our roots are weakened.

Literally. I had had no immediate expectations of buying a place of my own right away. In Quebec, where renting was a cultural custom, I had always been a renter. But in coming home, close to my heart, I carried the thought that I would eventually be able to buy my first home, a little adobe. Now, the extravagant wealth in real estate was more than a hindrance to this dream of putting down real roots. Struggling, scrambling, shuffling to avoid more extreme poverty and deeply hurt by the exclusion the new economics imposed on my life, I avoided the Plaza, where shops and galleries catered to the well-moneyed and their image.

Eventually I would understand that the same appropriation of local culture would impact Montana, Oregon, Alberta, and British Columbia, but for now I was busy, soon dismayed, sometimes bitter, living in a poverty rarely acknowledged in North America. Without the sense of an intact community, my spirit was waning while outsiders, often brandishing newly discovered Stetsons, proudly bought up the land and Pueblo pottery. The tacit understandings of the beauty and the demands of the land itself, the necessity of our survival with it, and the pleasure of cooperation among people, all values I had grown up with, were under duress. I took to writing to save my life, staying alive looking for words and shapes and forms that would reveal what "human" means in an impoverished culture. But still, almost desperately committed to living in my land again and trusting that, as they always had in my previous life, things would work out, I chose, again, to stay. Little did I know, as I resigned from my teaching position in Quebec, that the past does not necessarily repeat itself. Before this reality would settle in, I read these words from Lewis Hyde's *The Gift: Imagination and the Erotic Life of Property*. I heard them ring, recognizing a saving grace:

> *Zoë*-life is the unbroken thread, the spirit that survives the destruction of its vessels. But here we must add that *zoë*-life may be lost as well when there is wholesale destruction of its vehicles. The spirit of community or collective can be wiped out; tradition can be destroyed. We tend to think of genocide as the physical destruction of a race or a group, but the term may be aptly expanded to include the obliteration of the genius of the group, the killing of its creative spirit through the destruction, debasement, or silencing of its art. . . . Those parts of our being that extend beyond the individual ego cannot survive unless they can be constantly articulated. And there are individuals—all of us, I would say, but men

and women of spiritual and artistic temperament in particular—who cannot survive, either, unless the symbols of *zoë*-life circulate among us as a commonwealth.

Zoë-life, the spirit life. The loss of this commonwealth was the phenomenon underway in my homeland. The land had been at once the ground and symbol of our culture—indeed, the actual matter of its art among the Pueblo, Navajo, and Hispanic peoples. I feared the perishing of the homeland whose cells of aspen gold, bright streams, and vast light enlivened my own. A returnee, I realized I also belonged to the new graftings in this society where many have suffered the personal loss of cultural moorings. I feared for my own perishing. I am still deeply saddened when I see the loss of the graciousness of the bonds of humanity whose rules were respect of solitude in the land; cooperation or "pitching in," as my grandfather Reddin called it; and unquestioned generosity, called "hospitality," that made up the western way of life—invisible to most easterners who saw only the "cow town" and violence often (yes, it's true) engendered by the harsh conditions but stereotyped in the cowboy, Indian, or animal ruthlessness pictured in films. I am chagrined to think that drinking in the sounds of the stream, accompanying the silence of the grazing elk, or encountering the surprise of nesting osprey high above the trail may not exist for the younger generations exploring newly designed well-manicured nature trails and living museums. In spite of the educational value these managed environments bring, time there is most often controlled and codified, distancing people from their own learning and resonance with the realities of the natural world. Time, our proverb says, is of the essence. Time to walk, to listen, to see, to come upon new understanding, to wait for it. To learn to be careful. Time to wonder.

TWO

Whose "Wasteland"?

By 1803, the year of the Louisiana Purchase, fur trappers had already coined the expression "The Great American Wasteland" for Texas and parts north of what was still Mexico, where the Spanish prohibited their trade. This perception of a useless land remained in the American culture, sticking during the colonization of the West. Along the Rocky Mountain/Río Grande corridor, it long afforded us—descendants of the Native peoples, the Spanish, and the Anglo-Americans—protection against industrialization and urban development. But those days are gone. The "Wasteland" has become desirable.

One hundred years after the Gold Rush, the arrival of the railroad, and the establishing of reservations, history brought a new phenomenon to the corridor and the West as a whole. The technological consumer society imposed its greed and ways in an enthusiastic discovery, desiring its own sidewalks, condos, water, and written rules in the desert, where it has seen a nothing, or rather, a nowhere of dust and hills that could be turned into property in a relatively uninhabited land, no longer free for the taking but certainly for the taking. Where does this appetite for owning land come from? Ignorance? Certainly. Exoticism? Romanticism? Plain hunger for power?

Sitting with a friend while watching the moon rise over a rim in Canyon de Chelly, riding horseback with Mom in the Rockies, or alone on the road watching the northern lights dance between Montreal and Quebec, I've always loved singing our old western song "Don't Fence Me In," its cobbling rhythms echoing our love of moving through our open country. As the Inuit in the great expanse of the far North teach us, there is no "nowhere." A *here*, a *where* in the land, is always somewhere, never empty—not just property but a land alive with its winds, ice, and rains; its deer, geese, and lichen; its peoples' movement and settling in rapport with them; an inhabited land that makes us who we are. Unaware of this, insensitive to the meaningful ways and goings of animals and humans already in place, the impetus to buy and spend millions, an insatiable appetite, can take over. What does the appetite want? Extreme beauty, endless space, unspoiled land? These are understandable desires, not necessarily contradictory with westerners' culture. We too are proud of the beauty of this vast sky and earth resonant with our memory; we nourish our souls in ritual with the sage and our health with the green chile the land gives. The rocks and rivers accompany us like family. My own cells awaken with a sense of being kin to the blue spruce or the muttering waters of Bear Creek or the Bowen River. Oh but how I resist that appetite's pretension! The pretension to owning it all and, worse, the actual ability to own so much of it without knowing or often caring what it means.

A wealthy Texan friend, seeing my anger at the changes wrought by newcomers in Santa Fe, exclaimed one day, "My, you really don't like money, do you!" "Actually," I replied, flushed and smiling, "I like money!" I wanted more of it. I was spending my days scrambling and scraping to find enough of it. "But seriously, greed is another matter. As death is not murder, having money, being prosperous, is not greed."

What I mean by greed goes like this: with the recent influx of Santa Fe–style architecture into the mainstream, one couple arrived in a sparsely populated village an hour outside of Santa Fe. After buying several acres of land from an elderly Hispanic farmer, they built themselves a beautiful $10 million Santa Fe–style adobe home. The guesthouse was built for $2 million. In spite of their building a guesthouse, they clearly did not grasp New Mexico's generous tradition of hospitality. Once settled in their new home, the couple proceeded to sue the elderly farmer because the trailer his family lived in (as many people do in New Mexico) was an eyesore. This was the new way of the land.

Elsewhere, the statistics Joan Didion cites in her book *Where I Was From*, reveal the trials of southern California's urban economy when, as of the late 1980s, Wall Street's bull market was affecting the whole country. She reports:

> According to the Commission on State Finance in Sacramento . . . some eight hundred thousand jobs were lost in California between 1988 and 1993. More than half the jobs were lost in Los Angeles County. Before 1991 ended, California had lost sixty thousand aerospace jobs. The Bank of America estimated six to eight hundred thousand jobs lost between 1990 and 1993, projecting four to five thousand more . . . in the state's "downsizing industries" between 1993 and 1995.

In 1993 she remembers, "A well-known residential real estate broker on the west side of Los Angeles was advising clients that the market in Beverly Hills was down 47.5 percent." The boom had gone out of California. "The money," Didion succinctly states, "had gone away." in real estate, aerospace, and defense, and banks were letting go of thousands more; businesses were failing. 1988–1993. Crucial years in southern California, crucial years for me finding my footing, crucial years for Santa Fe. Years when New Mexicans were already anxious, dealing with a new "discovery" of their land. The unsettled economy in the Los Angeles area brought a large number of homeowners to the Land of Enchantment, where they found cheaper property values and lower taxes conducive to maintaining their high-end lifestyles.

Among "them" was a woman, probably in her thirties, who bought a home in the adobe compound where I lived toward the end of my first few years there. This was the last Hispanic neighborhood in the Canyon Road area, where I had found a rare, reasonable rent for a tiny apartment in the pleasantly rambling adobe enclave. The larger house this new neighbor bought, directly across from my place, had been owned and renovated by an old-timer Anglo, the only house of its kind in the compound, where a dirt road ran down the middle. Here, the relationship with the neighbors was the comfortable western one I enjoyed: one of daily warm greetings, respect for one another's solitude, shared delight in the rain, mutual readiness to help when needed. One morning while I was sitting just outside my window to get some sun, the new woman arrived in her Cherokee, the "must" vehicle for her class at the time. Since the road was narrow, she had to drive about four feet from me. Seeing her approach, I waved in the usual country way, and, as if on cue, suddenly, nervously, she rolled up her window and looked rigidly away then ahead as she navigated a sharp turn into her parking space directly across the way. No greeting. I was clearly poor—was she afraid of me? Was I not to be spoken to? Maybe dangerous? And at such close proximity, practically in her front yard!

What a discreet and quiet thing waving is, and how important in a culture that values its greeting. The absence of it that day was like someone turning a cold, deaf ear to the unobtrusive flow of human presence in our simple compound, cutting off that daily recognition of who and where we are. Perhaps being unsettled, she didn't know where she was. As an American and a westerner, I would have expected to find common feeling with this neighbor and her fellows, but instead it was lacking.

A century before, the journalist Charles F. Lummis had found common ground in New Mexico, having left Los Angeles to recover his health from 1888 to 1893. Quickly at ease in the saddle and befriended by the people of Isleta Pueblo, he recovered and became a defender of Indian rights. On returning to California he wrote *The Land of Poco Tiempo*, which sparked a literary movement bringing California artists and intellectuals to the high-altitude mesas and mountains well into the 1920s. In contrast to their own predecessors and wealthy eastern strangers-turned-neighbors who had embraced the rigors, health, and freedoms of the West since the mid-nineteenth century, a willing interest in the Southwest's peoples was not always what drew the more recent Los Angeles migrants, mostly of my generation, to the Southwest. Reestablishing their economic advantage in real estate, claiming the place and its beauties as their own expensive commodities, seemed more often to be at the heart of their identity with the place they'd come to.

This wasn't true of every individual. My doctor, who had arrived from Los Angeles with his family years before, had clearly established a natural and unpretentious rapport with his neighbors and the local cultures. From Northern Californians who related more easily to the meaning of the land, I learned there was a difference in values in the Golden State between their north and their south where image and entertainment dominated. A good number of the new arrivals ignored values other than consumption and appropriated the rambling adobe homes and court-yards, the Pueblo pots above the kiva fireplace, the Stetson hats and Navajo turquoise, the long views over the land bought and often put beyond reach of the people who created them. Familiar western gracious-ness chafed under the modern-day colonization, and more Hispanics abandoned their neighborhoods as the extravagant newcomers moved in. Our zoë-life, the collective spirit among the larger community, was muted, perverted. The attraction and arrogance of greed was muddling our sense of the sacred. I felt stifled without it. My diary received the truth of my fear and frustrations.

> Fran, tried and true artist, rooted New Mexican, told me this could happen. I feel less and less whole, less and less sensual, less and less safe, less and less believing. More constantly defensive. More mistrusting. This is new to me. Alienation is a different thing than suffering. I resent having to be or look like I'm rich to be valued. I hate this society for that. I fear becoming defensively pretentious here, all fake, no substance. The big danger.

What had happened, I asked, to the recognition of the quality of the person rather than her royal trappings that early Americans strove to establish? For the first time, close up, I was witnessing it being severed from the value of human reciprocity, care, and accomplishment. Fortu-nately, a man or woman's quality was still welcomed among the residents with the spirit of the place, for whom the difference in wealth was not paramount. As instructive as my experience might be about our Ameri-can values, it was painful to see too much money, this new facet of the society, maintaining its privilege at great cost to the local culture. But, I stopped myself, could it be that behind the presumptuous image some of "them" were on a simpler journey, perhaps similar to my own, hoping to recover a relationship with the wild? Or themselves? To be at home in the land?

Having been an immigrant, a foreigner, for years in the north, I was familiar with being perceived as an outsider in a minority culture whose language was threatened. In spite of the friendly curiosity of Quebec's people in general, in spite of their deep tradition of hospitality so similar to that of westerners, in spite of hundreds of years of cohabitation and intermarriage since British colonization, Quebecois could sometimes perceive me as one of the colonizers: an Anglo (here meaning an English Canadian) who was mistrusted and often ignorant about Quebec's realities. In spite of the exceedingly supportive and close-knit Quebecois community I was part of, in spite of Quebec's larger community embracing me as I guided thousands of students on their path to the future, the stereotype could take over and I could experience the resentment of their culture toward Anglos or—the sting at the other end of the stick— when speaking French in a café or airport, the contempt of an American or English Canadian for the Quebecois. Now I felt resentment toward the newcomers here.

I questioned my own tendencies to stereotype, but as one shop owner I often conversed with reminded me, I was not "making it up." It was no secret that up on the hills or along the acequias, locked gates protected a new, separate class, a late-twentieth-century conformity. Not so much a community but a class of buyers adhering to one image: Santa Fe style. This display of affluence, of purchasing power, this identity with it, was not what had been at the heart of the West I wanted to come home to. It was so far from fulfilling a constitutional ideal—not at all nurturing an established e pluribus unum but eroding it. Even years later,

Melissa Pierson's book title rings deep: *The Place You Love Is Gone*. Was it gone? Or just lost to me? The wrought-iron gates being built around the million-dollar homes belied the breaking and straining of trust among a people with a history of communal lands, an affront to neighbors who still looked out for and took care of one another independently of economic status.

Aware of my pain at the changes in town, a musician friend and his wife kindly offered me some respite, house-sitting their farm while they went to the Grand Canyon with their two boys. Between caring for their vegetable garden and greenhouse plants in the plains east of Santa Fe, I came across a nicely worn copy of *Democracy in America*, a still popular staple for thinking about America, written by the French philosopher Alexis de Tocqueville in 1835, after his extensive travels in the United States. Still grappling with understanding my relationship to my culture, I engrossed myself in his insightful discussion of our country, the "image of democracy itself." Speaking of the revolutionary Anglo-American democracy he respected and so keenly admired, de Tocqueville states, "I know of no country, indeed, where the love of money has taken stronger hold on the affections of men."

This portrait was beginning to fit. Not part of the United States in 1830, the Southwest had been a different country, not totally free from greed or meanness, but grown of different forces. The rich culture of its peoples had created a region in the United States where one still can enjoy earth-based values and the spirit of community they allow. In spite of the distinct identities and painful historical conflicts among the Indian,

Hispanic, and Anglo peoples, the tapestry of history and customs has led New Mexicans and other southwesterners to bond through collective values, nourished and supported by their traditions replete with celebration of the cultural and spiritual sense of the land.

For years, the traditional barter system had allowed New Mexicans to survive in a desert agrarian community. Men and women coming home from World War II ushered in a cash economy for the first time, but the old ways of trade held. Still imbued with the spirit of independence and cooperation that barter in a survival culture allows, they held again in the 1960s, when southern California's burgeoning agricultural industry usurped New Mexico's regional agricultural economy. The spirit held for a few more decades until displaced wealthy newcomers entered the "Wasteland."

Historically land-rich and cash-poor, now in the mid to late 1980s, unable to afford their own homes, Santa Feans in greater numbers had to uproot, moving to Albuquerque, Denver, or Californian cities for work. In 1991, Santa Fe, a Spanish-speaking capital since its founding in 1610, became for the first time majority Anglo. A mercantile spirit from somewhere else in the United States had begun to dominate, ironically, in a kind of modern reconquest from the West Coast rather than the East. In the meantime, in the northern section of town where I worked and sculpted, new zoning laws proposed by the newcomers were in the making, changing established understandings of what neighborliness meant. Under the dismembering and dismantling effect of the new wealth on the fragile local economy, the barter system as a viable economic structure dwindled, then died. Newfoundlanders way up north in the Atlantic know this story too, as do residents of rural England, a British friend tells me.

House-sitting had long been part of the Southwest's barter economy. When I arrived to settle in 1988, I had counted on this tradition and found it still viable but fading. This was a custom whereby wealthy New Yorkers or Texans who had second homes in New Mexico would ask house-poor local people, often Anglo-American and well educated, to live in and take care of their homes. In this mutually useful and satisfying relationship, it was the respect between owner and house sitter—the quality of the person, not the wealth—that mattered, reflecting a truly western arrangement in a familiar Western Spirit. Based on one's word rather than written agreement, the trust between individuals was rooted in their common bond of love for New Mexico, the land itself, inseparable from its people and their customs. With the assistance of my brother-in-law, Paul, who was always ready to help me move, I enjoyed two years of the old way. In a state with the continent's greatest number of PhD construction workers, this tradition had guaranteed long life in the Southwest to many an inhabitant.

With the changing population and absentee ownership becoming the rule in Santa Fe, this custom of house-sitting transformed now from a cooperative caretaking of land and home to a new class-conscious, cash-based arrangement. By the time I would leave in 1993, a new generation of absentee owners was charging house sitters to take care of their large homes. Unthinkable even ten years before. A normal practice, though, in a different American way. But customs are resilient, people tenacious, and third and fourth homes do need a human presence. House-sitting continues, occasionally still in tune with the old spirit of barter, other times – can we speak of a diversified economy?—with individuals hired to housesit or some still asked to pay for taking on the responsibility.

As the end of the twentieth century approached, the way of life had changed for all Northern New Mexicans—Hispanic, Anglo (in this context, meaning anyone who is not Hispanic or Indian), and Native peoples. There was much cultural self-consciousness with many individuals scrambling to "get a grab" on the *nuevos ricos*, "the new rich," all the while resenting their economic power.

I awakened one February morning, aware, knowing that the place I loved was gone. The place of familiar customs and tacit reciprocal bonds—its community knowing how to be, to be free to roam, to both care for and let be the waters and desert, to fight but survive together in the high, dry land—was gone. The new way not familiar as the way of home. But wait, perhaps not. It was not really gone, just becoming a subculture in its own territory.

In the interim, house-sitting and my multiple low-paying jobs had not allowed me to buy a house or rent a safe place in the newly high-end environment. I held on, continuing to paint and sculpt in the slices of time in the studio, but I too would soon be gone. When a bout with meningitis paralyzed me, it would convince me the time had come to go back to Montreal to find financial health again. I would trust an intuition that a teaching job would be waiting for me. As I recuperated, walking along the river by the studios, admitting I was fortunate in spite of my grief to have a choice, making promises to spend winters in my homeland when I retired, I was unable to find spiritual or emotional meaning in the departure itself. Before accepting the inevitable reality, a painting surfaced with the title *Variations on Undesired Flight*. For a while, I wrote in the refuge of the studio. My poetry spoke of the loss.

Leaving Home

I found this West
living now on its own peripheries
searching out refuge
the center place of its wide simplicity huddled,
retreated into people's hearts and memory.
I found this West straining under the grafting of greed
and its companions, suspicion, envy, and denial.
Santa Fe. Central City.
I found this West in name and image
with very little likeness to itself.
I found this West losing ground.

But the beauty of the land held. One evening an old-timer New Mexican (who no longer lived in the state) remarked during one of his visits that "what's here—the place, the desert, the mountains, the vast sky, the petroglyphs—is bigger than all the greed, all the loss." The land and its spirit had always been our root, what mattered. Yes, I thought. In this wild that remains where I hike, walk, paddle—in this land every tree, color of blue, every grasshopper is exquisitely familiar. I am at ease. I know them, they know me. We speak, the silent deer, the coiled snake, the playful raven, the water and I. I recognize the dangers and do not interfere. This gift of "at homeness" I have always had with the light, the shimmer of the aspen leaves, the sound of silence alive with inaudible life and the scurrying of some. They are in my cells. A gift of my birth, a gift of this West.

Until "the mountains turn to dust" we have a home, my eldest sister Mary had said. We are not lost yet, I told myself. Given the changing economics of the land, though, there was no house or home to call my own. No place to rest, to gather and continue celebrating with my fellows, knowing together what this place means. I was reverting to being a nomad. The land and the sky will hold on. But is it only the very wealthy who can live here under the blazing blue? What does it mean if we can't live in the land, if we barricade our homes against one another? This is not, and never was, the western way.

THREE

The Mobile

The sense of a mobile society's uprootedness was not strong during my childhood. The movement between our tree-filled neighborhood in town on the flat Denver plain and the family homesteading cabin in the mountain wilds not far away was part of the adventure but also part of the settled character of our lives. The stories of family history were sparse. Our grandparents' worlds before the Thanksgiving meals they served us, the rigors of the Taylor grandparents' lives in Colorado Springs and of the Reddin grandparents' adventures in Denver, were hard to grasp. The Klu Klux Klan attacks against Catholics and a violent miners' strike in Colorado Springs, destroying my paternal grandfather's craft as a linotype printer along with the newspaper's presses, were exciting, intriguing stories, but distant.

It was 1958 when our dad took Kay and me to a historical village the city of Denver had built, a replica of a Gold Rush town that had mushroomed on the Platte River one hundred years before. Denver, the state capital, was celebrating its first discovery of gold and the early influx of American prospectors and settlers into the then territory. Colorado's past. A new population, a new way of life beginning then. I remember the somewhat nondescript cabins and, probably because of my interest in calf roping, asking my dad a few questions about rope making. The historical set, a main street, was lifeless but not foreign, easy to compare to the log cabins and lodges we knew in our mountains. The fabricated street was similar to those of the old, small mountain towns we drove through on the way to the family cabin near Grand Lake. In a sweeping valley near Granby, there was Fraser with its seven residents, a town that intrigued us as the coldest spot in the nation, evoking a strange kind of pride in the rough mountain life. Among its four or five wooden buildings was an identifiable post office, a dark storefront. A slight smell of horse manure gave witness to the fact that lives went on though the place reeked, of both the manure and the neglect. The familiar architecture we saw in Denver's downtown setting gave me no more than a child's vague grasp of what the replica of the Gold Rush days—and perhaps the West—were supposed to mean.

My real sense of those miners' lives came from Dad showing us children the remains of their cabins in the high-altitude forest as we foraged for wild strawberries and, in rare years, blueberries. Gaskill gold mine, a few hours' hike above our log cabin, made history real, especially when finding remnants of rusted mattress springs and, just outside our cabin's front door, old tins from the saloon in a hollow that had been used as a dump.

History became more complex later when, as a twenty-five-year-old, I read the story of Geronimo written by a man who had been in the Washington, DC, administration at the time of the government's attacks on and capture of the resistant Apache leader. Wanting to tell the truth—one of the first to do so—the man not only was straightforward in his indictment of the government but used honest language that allowed his human concern to come through, making the injustices toward the Indian people real, and shocking. That summer I had already been living in Quebec, where over three hundred years of European history was the norm. Vacationing at my parents' house in Denver while reading this man's witness, it struck me how young American history was in my homeland. Less than one hundred fifty years, starting with the subjugation of the area's Indian peoples: Navajo, Apache, Ute, Pueblo. The miners' strike and the Klu Klux Klan agitation had a larger context now, part of the previously unsettled, mobile American society getting its footing in the West.

Freedom demands space, place, and land, as all Americans know, as well the Native nations whose range of freedom for hunting and ceremony were wrought from them know, as does the government that tried to subdue them. Transient strains of a developing society as well as nomadic traditions make up our history. The Gold Rush is a reminder that the American West has long had a highly mobile population. While the Pueblo are village people, the Navajo, Ute, and Apache of the region were nomadic or semi-nomadic before the Indian Wars and the reservations that followed. Like the Apache from the plains, the Comanche raided and traded with the pueblos. The Spanish explored and raided before founding Santa Fe in 1610. Although destructive to the Native communities, major Spanish incursions were initially short-lived, due to the absence of gold and the resistance of the Pueblo people in the Revolt of 1680. Scattered Spanish families were left to fend for themselves without the protection of the Crown, founding their own farms and joining both in marriage and in battle—at times allies, at times enemies—with the Native population. General Diego de Vargas's reconquest for Spain in 1692 brought both upheaval and eventual reconciliation between the Spanish and Native peoples.

Early on, the Comanche traded and raided in Taos from their sanctuary along the Colorado plains, and other displaced Plains peoples occasionally crossed the Rockies for the same purposes. Europeans like the explorer John Charles Fremont set up expeditions penetrating farther West. The end of the Mexican-American War, sometimes called the War with Texas, brought the Treaty of Guadalupe Hildago, which increased overland traffic on the Santa Fe Trail and facilitated the dreams and ambitions of new territory among Indian agents, soldiers, lawyers, missionaries, settlers, bandits, and merchants from east of the Missouri River.

Among the young Americans to ride over the Santa Fe Trail were adventurers like Kit Carson. Yes, the man the Taos elder had spoken of. Linguistically gifted, first a friend to many tribes and fur traders, Carson eventually, albeit some historians claim, reluctantly, led the assault on the Navajo for the American Army, leading to the burning of their fruit trees (the first "scorched earth" policy) and the people's subsequent starvation, smallpox, and too many deaths on their forced Long Walk to confinement at Fort Sumner. A devastating story and only one of the many true stories of the uprooting and exiling of the Native peoples in their own land.

Greed is an ancient human failing. It scathes the heart to realize how deep it runs in Europeans' relationship with our continent. In reading Frederick Turner's *Beyond Geography: The Western Spirit Against the Wilderness*, I fell upon the scene where he reports on Hernán Cortés's encounter with Montezuma's ambassadors who had come with a "plentiful display of gold . . . from their great lord" to discover what "manner of men these [Spanish] might be. . . . "Send me some of it [the gold]," Cortés urged, "because I and my companions suffer from a disease of the heart which can be cured only with gold." This rings too true nowadays. Yes, my heart, greed is a brutal human failing.

A great civil war tore asunder the eastern United States in its relationship with the South as the struggle to release black people from the bondage of slavery marked the continent's history. However, in the "Wasteland," while the mining resources in the new territories of New Mexico (1850) and Colorado (1861) were something the confederate South coveted, the small population in New Mexico was maintaining a desert survival culture that allowed for distinct customs and communal uses of the land. In this land apart, the debate over slavery on the southern plantations was not paramount. The distinct societies of the Southwest understandably felt geographically and culturally distant from the economics of the Anglo-American way in spite of their own complex customs of taking captives, which did sometimes involve making people slaves, notably among the Comanche and the Spanish.

Anglo-Americans had been trickling in and blending in since the early days of the Santa Fe Trail, first opened by Mexico in 1821. Influential as merchants and lawyers, they were few in number before the Civil War, and the Pueblo and Hispanic peoples continued their hunting and trading with little disruption, thanks in part, the historian Marc Simmons relates, to the protection of the Comanche. Once the Confederate attempt to take New Mexico quickly failed, the peoples of the dry territories of Colorado and New Mexico were left alone, for a while yet, by the falsely proud "Government of a white race," to quote Senator John C. Calhoun's famous phrase. Until, that is, after the Civil War when the United States Army found new direction, conducting the Indian Wars against the Apache and Navajo (both nomadic tribes) while the Pueblo suffered less displacement but extreme neglect by the new government. An agricultural people, they had been recognized as citizens by Spain and, protected by the Treaty of Guadalupe Hildago in 1848, granted full citizenship by the United States. A status they still enjoy today. Although over time they have to fight to reassert their rights, guarantees to their land and property rights were upheld during the upheavals of the late nineteenth century. The Ute and the far-ranging nomadic Cheyenne, Arapahoe, Comanche, and Kiowa were part of the movement over this land too. Trade, dance, and other ceremonies continued, as did hunting, raiding, and the newer activity of conducting war parties along the Santa Fe Trail.

The history of this land is nothing but complex! The abolishment of slavery as an economic system in the United States is a historical given, a social triumph to be proud of, but the ensuing takeover of land in the West broke spirits, killed and subdued whole civilizations indigenous to the continent, and established reservations where the majority of Native peoples, the Pueblo being the exception, were considered wards of the state.

These were the wars in which Kit Carson had accepted to lead raids on Navajo crops and farms for the US Army. These were the wars that the Taos man spoke of to that ten-year-old child when she visited his pueblo. Had her kind father been aware that it was on our own soil that the American General William Tecumseh Sherman had established the prerogative of the US. Army to invade and kill people without just cause? The stories about the genocides in the West are legion, best told, I believe, by the Native people themselves. But I sometimes wonder if the good father who taught the ten-year-old about the United States had known that it was the Washington, DC, government itself that gave General Sherman, supreme commander of the armies in the West, the power, the permission, the right for his unjustified seizure of territory and massacre of the Sioux?

Although we are gradually becoming better informed, there is still much deep-seated ignorance of our own continent's history of hate. My father must have known of it since as a young man he had worked a short time among the Oglala Sioux. Perhaps he could not overcome the fear of the devil that his Catholic tradition had given him in its attitudes toward Native peoples. He taught us a few words in Sioux, which I would confuse with the few words he taught us in Italian, but never did he speak of these wars on the Indians. Although he did recount a story of fatigue saving him from opening a window where a bomb had been set, and another where he was saved from drowning in the Mediterranean—he was convinced—by Mother's love via a medal she had sent him. Like many of his generation, he never spoke of the horrors of the Mediterranean theater of war he fought in. He would speak, though, of the blue of the Mediterranean sea, the beauty of Italy, the kindness of the people. In a postcard he sent from Palermo in 1943 to my sister Carol, he wrote to his two-and-a-half-year-old, "Darling Carol Anne, please save this card for Daddy. It shows the water and the mountains. Aren't they pretty? Daddy." He doesn't mention the warship visible in the distance. The censor would surely not have permitted it, but the omission was consistent with his way.

When older, growing into social awareness and interested in such things, I might ask questions of him. His response was to remind me of the musical *South Pacific*, in which a song tells how hate and fear have to be taught, that to work they have to be repeated, harped upon. This kind of teaching he did not drum into us, but worked quietly and successfully for justice, facilitating legislation for Amish home education rights and Catholic children's access to schoolbooks. My parents had not taught us about slavery, and I heard him, though only once, impatiently use the word nigger, but my father worked indefatigably, often pro bono for a good part of six years, in defense of the black pilot Marlon D. Green, bringing his case of discrimination before the US Supreme Court. My father's argument relating to interstate commerce won a unanimous decision in Mr. Green's favor, providing the key for passage of the Civil Rights Act of 1964, as well as making Mr. Green the first African American commercial pilot. I was fourteen at the time. Dad had flown my mother and all of us children still living at home with him to Washington, DC. I continue to enjoy the memory of that, my first plane ride. And how could I not be impressed when Bobby Kennedy, then US Attorney General, came and sat directly in front of us to hear my father plead before the impressive Court? But what I remember more fully was a moment in his downtown Denver office where my sister Kay and I delighted in visiting him in his separate world. This Saturday morning, perhaps a week or two before we left, I remember sitting very quietly near the wood bookcases, absorbed in reading the brief he had prepared for the case. I learned that it was the one hundredth anniversary of the liberation of slaves in the South. In the preamble to his argument, my father invoked Abraham Lincoln and that anniversary in presenting his moving legal plea for justice in Mr. Green's case. That Saturday I believed for the first time that good work could actually heal human history. Knowing his commitments leads me to believe now that if this quiet, dedicated man knew more of the horrors and griefs of our plains and mountains, he chose not to teach America's story of hate and greed to his children. There was another kind of world to be handed down.

Although Sitting Bull was still alive when my maternal grandfather John Reddin arrived in Colorado in the 1880s, the Indian Wars waged by the US government had changed the face of the West. My ancestor must surely have known well the stories of actual unbridled freedom of the invaders and their army. In the Southwest, Geronimo gave up his long resistance in 1886. The Navajo had returned from the trials of the Long Walk to their traditional lands among the four sacred mountains, but their home had been declared a reservation now. The Comanche had made peace and stopped their raids on the Navajo and Spanish. The Ute Chief Ouray, born in Taos but well-loved in Colorado, had just died. Gold and silver rushes, the railroad, the wars. What my grandfather knew or thought about this huge thrust of development and change through money, war, and migration in this West he came to love has not been handed down to us. He was part of the new migrations. He shared in the rare letters to his children only the joys of the wild. Greed may be part of the root at the heart of our story, of our relationship with this vast land of America—part of it, but not the driving force of an inclusive e pluribus unum, which would involve cooperation in caring for community and the land.

During the Civil War, the Confederates retreated to Texas after being stopped at Glorieta Pass, New Mexico, in 1862. The route toward Central City, Colorado, and the much-coveted Colorado mines remained permanently out of their reach. The sparse population of the Rocky Mountain corridor and its desert plains continued their lives a bit apart from the worst ravages of the conflict between the North and South, a major trial for the United States and its principle of e pluribus unum. That is not to idealize the frontier. A transient place, it could be dangerous too. You had to hold your own, know who you were, and how to survive. Even if the independent, quieter ways of Southwest life went on, consequences of the slaughter of the buffalo, cattle feuds, the gold rush, and raids by outlaws and bandits became part of its fabric. There is both joy and meanness to this West.

At the time Grandfather Reddin, a lawyer, established the Knights of Columbus in Denver in 1902, there were very few black people in Colorado. As a result, Catholics became the targets of the Klu Klux Klan. Today my parents' graves lie in view of Table Mountain, the mesa where my grandfather had been hung and burned in effigy by the Klan. The land carries all the memories we give it. It is comforting somehow, a century later, that their deaths were peaceful, comforting to know they repose in the land they loved, that my mother and father survived that history. A continuity that remains alive for us, their children. The unsettled West of my grandparents' time in the Rocky Mountain/Río Grande corridor had eventually settled, allowing small populations of Anglo-Americans to homestead, root, and join the Indian and Hispanic peoples in surviving in affection for the land, passing their reverence for its beauty and power on to the children who would be its new inhabitants. Trading never lagged; mines boomed, died, and operated again; farming continued; ranching developed; and towns grew. The Puebloans and Navajo never ceased their lives of trade and ceremony.

Mobility and transiency are of a different order now, although still on a continental scale. They may not be an American invention, or new to the continent, but in the modern West it seems mobility has become a norm with individual wealth as its motor. Naturally, we can only be grateful today's movement is less bloody. The duress, however, is not necessarily less profound. Absentee owners jet from Montana to Los

Angeles, to Santa Fe, to New York visiting their third, fourth, or fifth homes. A phenomenon now familiar in the beautiful places of the West. But this new mobility does not necessarily lead to an ability to root, to relate to place and community, to have or recapture an identity. A kind of continuity with the old-fashioned transiency is broken by the imposition of a new class whose wealth does not lead it to integrate culturally, and where there is no "trickle-down effect" to allow the local community to hold its own economically. The real estate boom for the affluent has continued in Santa Fe, accelerated in Taos, and burgeoned farther north along the Rocky Mountain Front Range in Colorado. In Taos, where, according to local leaders, the average income for a working person in 2007 was $15,000, the average home cost $334,000. Interestingly, in 2000 the US Census Bureau had listed the per capita income at $15,983, climbing in 2009 to $22,868, which was still only half the nation's average. In Santa Fe, the average home in the spring of 2007 cost $517,000. The Census Bureau identified the per capita income there as $24,967 in 2000, up in 2009 to $30,714. And poverty remained. In the 2000 census, the percentage of people living below the poverty line in Santa Fe ranged from 12.3 to 22.4, depending on whether they were family or single-parent households. In Taos the range was 21.2 percent to 33 percent.

More displacement threatens our culture's coherence. The anguish of many an American is a loss of identity due to the loss of community and geographic relationship. Following the loss of intergenerational relationships with place—especially intimacy with its animals and plants, the "biotic community," as we now say—the possibility of forging a sustainable identity has been looking bleak for westerners. So bleak that ranchers have been fearing the loss of their way of life since the 1970s.

Ownership of land under the thrust of greed, whether individual, corporate, or government, displaces established communities, leaving people impoverished in their ability to maintain identity and generate kinship with a place for themselves and their children. The dislocation of inhabitants leaves the land without care and creates the holes in the culture where violence is easily generated in place of belonging. Modern-day bureaucracy may seek to fill the gap but often succeeds only in trivializing local custom. As well the Native people know, when the bonds of belonging and custom disintegrate, the brutality of power—physical, economic, or both—accentuates. The land carries all the memories we give it. No wonder we carry a deep sense of our world itself perishing under the thrust of a narrow-minded economy putting strains on water and land, health of the forests and the animals. As displaced New Mexicans show through their resentment by vandalizing wealthy outsiders' cars, it becomes evident that the freedom to have a secure, settled economic relationship with our home region is jeopardized. Having the economic means to belong to, live in, and work on our own land was what we believed in the mid-twentieth century—yes, expected—that our American democracy was about. By 2008, we were far from Franklin D. Roosevelt's New Deal that mobilized Americans to improve their own infrastructure and environment in locally managed projects.

What are we to do when faced with the inevitable and irretrievable loss? Long a major source of water for New Mexico's biggest city, the aquifer deep in the land under Albuquerque, like many others in the West, is shrinking due to more and more asphalted roads, fewer and fewer farms and pastures. Under these conditions, neither the sparse rainfall nor limited irrigation of the soil can be counted on to replenish the groundwater feeding the aquifer, a moving underground river. There is, though, a project slated to transfer water from the narrow Río Grande to replenish the aquifer's groundwater. Such a transfer is problematic, however, as the Río's water is already overappropriated, and the continued building of new homes suggests little success. One hopeful sign is recent legislation to protect the Española Basin, a neighboring aquifer to the north, for that population's drinking water, its sole source.

In her book *Where I Was From*, Joan Didion, in dealing with photos and memorabilia after her mother's death, states, "There is no real way to deal with everything we lose." Perhaps. But we can witness it. And we can hold on to the real ways the land and the culture still offer. We mustn't abandon the values that gave us cohesive communities. The way of life that gave us those communities and the accompanying freedom to be alive in the wild that westerners have known is fading, constrained. The remaining land and watersheds need restoring and protecting—more than ever. Many children are wondering what we will make of their world, even while other children roll up their sleeves and pitch in. What we have known as the Western Spirit is worth remembering. With our living knowledge of the land, we were more human. Our humanity needs to be strengthened by it now in our complex times.

E pluribus unum was not the only value my parents gave us. There was the Western Spirit as well, resonant in the gentle wilds, and yes, on those days the mountains had called us from the city, when my mother and her daughters would ride alive over the frighteningly steep Fraser Pass. Together we sang, "She'll be coming 'round the mountain when she comes," as our mom spoke of her father who had ridden over this pass too, in "Mac's wagon" led by four horses. In his time it was a one-way trip of ten days just to get to Grand Lake, the valley town and lake our family still cherishes. A round trip, he recounts in a letter, required "thirty days of steady camping, fishing, and hunting," during which the jobs were distributed and "everybody was to be good-natured and pitch in." Here the risks of storm and wildcat were weathered, and the rare, tense trade of a lawyer's services for right-of-way through a defensive rancher's hunting territory was part of the adventure of survival. It is a paradox of the Western Spirit that the mean and gentle, the greedy and generous intermingle, and beyond their confrontation follows an under-standing. More typical on the range was the unquestioned warmth and comfort that the "old hermit, Uncle George" offered my grandfather and his friends in his cheery cabin during their long trek. Hospitality thrived in Taos and Santa Fe; at Santa Clara, Acoma, Santa Domingo, and Jemez Pueblos; and in Canyon de Chelly, among the Navajo. There had always been hospitality "on the range" where we lived, before the Longhorn and Hereford cattle grazed.

A youthful, wanting-to-be sophisticated teenager, I felt insulted one day when a visiting New Yorker told us our city of Denver was a "cow town." But it was true. Thank goodness! Roaming the land, hunting or fishing, knowing it; respecting the rhythms of the water and the move-ments of animals at dusk; extending hospitality to your neighbor, be she from near or far, New York, or even England; respecting her intimacy with the wild. These were basic basic tenets of the Western Spirit. "Cow town," indeed! My family was proud of my mom's cousin, Uncle Paul Patridge, and his blue-ribbon-winning cattle. My twin sister Kay and I loved the roundup we were once allowed to ride on his ranch. With Dad, we delighted in visiting the barns at Denver's annual National Western

Stock Show every year. But I inherited something more of the fabric of this worldview: the spirit and the knowledge of being born for joy, as the Inuit too believe, that comes with reverence for the land. This, I have found over many years, distinguishes my inheritance from that of many other Americans. Not the same thing as "the pursuit of happiness," it is a given. It demands reciprocity with nature.

Struggling in Santa Fe, feeling transient with my numerous jobs and no lodging of my own, I wondered: Has this failing of our culture, that deep thrust of greed, singled itself out and shrugged off all the other motivating values I grew up with? Unprotected in my returnee status, a different kind of migrant myself, I was getting to know the meanness of the West I had not known as a child, all the while reexperiencing its familiar generous spirit that did not translate economically.

FOUR

Influx

Browsing in a Santa Fe bookstore, I came across a postcard widely circulated in the Southwest though published in Montana—also a place of great natural beauty and earth-related culture that has experienced the arrival of an invasive wealthy class. The postcard depicts a rancher couple standing in a wide meadow near a lake, watching four flying saucers arriving over the mountain peaks and forest beyond. As the woman points at the airborne vehicles, the husband responds, "More Californians, no doubt." "Right on!" my friend exclaimed laughing heartily as the salesclerk and I joined in. Laughter is a saving grace, a relief to the frictions inherent in our efforts to adapt to the changes. Elsewhere, at the Elko Cowboy Poetry Festival, a crowd of westerners shared a belly laugh as the host's sharp humor captured how the culture spawns new rivalries with the old. "I grew up in southern Colorado," he quipped, "so I grew up around Texans. We used to think the most dangerous thing was a Texan with a high-powered gun. We were wrong. It's a Californian with a U-Haul!" Such a relief to laugh at danger, at loss.

Perhaps the reason the Los Angelenos seemed so alien along the Rocky Mountain/Río Grande corridor was that their own society had for a long time been "strictly man made," as Charlie Russell, the nineteenth-

century wrangler and famous painter, put it. Indeed, real estate development was a reality early on in Southern California. In 1920, at the time
homesteaders were arriving in the forests of Colorado, Charlie Russell,
retired from the range and on his first vacation to the Los Angeles area,
wrote to a friend: "This is beautiful country all right but it's strictly man
made. I think in early days it was picture country before the boosters
made real estate out of it but I'm about 100 years late . . . nature ain't
lived here for a long time and that's the old lady I'm looking for . . . but
there are still some wild spots." Over the last fifty years stories, poems,
and migrations have told us how much of Southern California's "old lady
nature" has been lost, in spite of the Colorado River water it uses.

Its history was shifting again in my lifetime. As they fled the strife of
their failing economy, the migrant wealthy of my generation adopted
New Mexico's capital—long called the "City Different" due to its Pueblo
Revival architecture—in the cultural mode they knew of ownership and
entertainment. Some embittered, long-established residents felt both
their neighborhoods and the Plaza were becoming a theme park, their
Disneyland. As earlier in the Colorado Rockies, and later in Montana and
the San Juan Islands, the unceasing increase in real estate costs did not
reflect an economic boom for the local people.

On the contrary, the radical incursion of outside wealth created a
profound change in New Mexico's traditional economy, affecting the way
of life in the capital and beyond. The sudden increase in property values
led to family choices like that of one of my Hispanic neighbors. At the
death of his grandfather, the family sold his home, a one-thousand-
square-foot "fixer-upper" for $1 million, the suddenly "usual price" in the
sought-after northeast section of town. From 1991 to 1995 in the City
Different as a whole, the cost of homes increased 15 percent while the

nation's average stayed below 5 percent. In 1985, a friend's sixteen-hundred-square-foot home in a pleasant semi-rural neighborhood only ten minutes by car from the Plaza was sold for $125,000. In the spring of 2008, in this neighborhood only somewhat less-sought-after than the Canyon Road area, her house would sell for $450,000, a 36 percent increase, just under the official average price of a Santa Fe home at the time.

The more recent phenomenon of a plethora of short-term rentals creates an even greater tear in the social fabric, because absentee owners can't be easily consulted when renters' behavior is disruptive. In this still small town, the mutual support among neighbors trying to protect the neighborhood peace, quiet, and congeniality is forcibly giving way to urban anonymity. One understands a friend's anguished dismay when she complains, "They're selling us. Santa Fe wasn't about the adobe walls but the way of life." The traditions of earth abodes, though, were integral to and symbolic of that way of life. There is a special beauty to the real adobe, the silk of hand-smoothed walls, the erotic, sensual peace of the earth about one, and the knowledge of the care that put it there, joining human hand, earth, and sky to contain the life within.

The years Joan Didion had identified for California's economic hardship, 1988 to 1993, were turning-point years for Santa Fe as well. From our various stances in the cultural fabric—some struggling, others at ease financially—friends, neighbors, store owners, and fellow workers acknowledged feeling, like myself, that these were the last years the old New Mexican ways held the social fabric together. Certainly the last years where real estate belonged to rural and working New Mexicans themselves. It is curious how a community knows, before the statisticians and politicians speak. The awareness expressed among us was like the shift of light on the mountains where the peaks suddenly and so quietly turn from blue to purple, and you know the day has come to an end. The subtle gift economy was being unraveled under the impact of the displaced wealthy. A world undone. Santa Fe was threatened with becoming more of a real estate market than a living, lived-in city, the Plaza more that feared shopping Disneyland than a viable town center. According to the US Census Population Estimates Bureau, the city counted a population of 49,100 in 1980. Since this initial period of influx, the struggle has continued as the number of people in 2007 reached over 73,000 in town and, with exurban development, close to 140,000 in Santa Fe County. The increase still testing, as one real estate agent put it, "how much money can be made on a mud house with wood beams?" Multimillion dollar part-time homes, short term-rentals, and "specs" continue to determine the living environment in Northern New Mexico in 2011. Incisions in the social fabric feel fatal to some. "It's over. Santa Fe is gone," people who grew up in the old ways and days are still telling me—but with more resignation than in previous years, when the loss was new and the sorrow fresh.

It seems from Charlie Russell's letter that the culture, "the genius of the group" that the essayist Lewis Hyde speaks of, can be lost as the land is lost. In the 1980s, in spite of their attraction to the beauty of the sparsely populated and often still-wild Rocky Mountain/Río Grande corridor, many among the newly displaced population in Santa Fe could not easily recognize the spirit and knowledge of a land-based community, no longer knew how to see the "land of the free" other than as a real estate commodity. Perhaps, coming from a strictly man-made place, perhaps in spite of having lost their own old lady nature to urban sprawl, this was all they knew when they came. It has not been so long since

nature lived in the homeland I speak of, but she is losing her footing along with that of her inhabitants.

The 1980s Santa Fe style led to more influx of mostly urban wealthy from elsewhere, too: Hollywood itself, of course, Chicago, Boston, New York. If Southern California's circumstances can be understood, Texas wealth—the other economic power driving accelerated development in the Rocky Mountain/Río Grande corridor—is another story.

Texan friends tell me they enjoy their notoriety as great braggarts. For myself and others who grew up in Colorado, it was common knowledge that Texans saw their land as bigger and better than ours. It might be offensive at times, but it didn't interfere with our laughing and teasing over accents and knowing the Texans as warm and generous-hearted. Mostly they are characters, I thought as a child. Gracious but different. This was familiar, like my redheaded, independent, and spirited Aunt Teetie, an artist and physiologist, who would "flyyy" up to Denver by car for the holidays from her home in Amarillo. The West was pretty much made up of characters. Their bragging was like Dad's being of English extraction and joking with an "All Hail to the Queen of England!" as my Irish mom wrote her St. Patrick's Day cards. In the American Colorado outpost, this was not offensive. Everyone had their quirks. Although the British did invest early on in the development of ranching in the West, Britain was no longer a rival—but Texas was. The bitter reality was there. Their ranches were bigger than ours, and richer. It was true.

Close neighbors and rivals, Texans are not newcomers to the economic and cultural fabric of New Mexico. They have loved this land—its wild mountains and desert beauty—for generations. When a Texan wore his bolo tie or turquoise, or adorned her fireplace with traditional clay pots, it was with a western pride and often knowledge that carried the warmth of the land. If New Mexicans and Coloradoans tend to be reticent to brag, even docile some say, whereas Texans like to show off, this difference in personality is not the root of their rivalry. Friendliness, neighborliness, a real value in both their cultures, overcomes this. It's economics that's the rub. Historically, and now.

Early in the nineteenth century, unrest was brewing in New Spain, which still included Texas, California, and the Rocky Mountain/Río Grande corridor. In 1803, the United States bought the vast neighboring territory of Louisiana from the French, doubling the young Republic's land mass, and guaranteeing itself access to New Orleans's commercial port and warehouses. West of there, the movement toward Mexican independence had already begun as Spaniards born in Mexico chafed under colonial rule and a famine caused by Europe's changing economics and this continent's as well. With the success of the Mexican Revolution in 1821, Spain, too, was forced to relinquish control in the New World just as France had been. In contrast to Spain, which had prohibited trade with the United States, Mexico opened its borders not only to fur traders but to American merchants traveling overland from Arrow Rock, Missouri, on the 1,203-mile-long Santa Fe Trail. By the time de Tocqueville arrived in New York in 1831, the small, still dusty town of Santa Fe, far to the west, had become a bustling center of trade.

In the same period, Anglo-Americans were migrating to Texas in search of inexpensive land. Under the new arrivals' influence, Texas became its own separate country. In 1836, its own revolution and declaration of independence from Mexico established it as a republic winding all the way up into what is now northwestern Wyoming, covering a good part of what fur traders and others had called the "Great American Wasteland." Nine years later, failing to gain recognition from Mexico as an independent state, Texas relinquished its status as a republic, accepting annexation to the United States. During these struggles, Texas had consistently fought for Santa Fe, the center of lucrative trade, to be its republic's northern boundary. Yes. The New Mexican capital has long held a place in the imagination and territorial ramblings of Texans.

As early as 1841, a group of Texans set out to claim part of New Mexico for the Lone Star Republic—unsuccessfully, as the New Mexicans' defense would have it. Five years later when the Americans, led by Brigadier General Stephen Watts Kearny, entered Santa Fe and claimed the alien (to them) New Mexico territory, it was saved from becoming an appendage of Texas. However, the fur trade was dying, the choicest pastures were already being taken over, and limited water precluded New Mexico from being an equal in the cattle industry upon which the new territories would rely.

To the north, the newly recognized Colorado range was invaded by Texas herds roaming free among the tall grasses in that same land where the antelope graze and where, so many years later, the meeting of plain and mountain would revive me on the drive to Santa Fe with Geoffrey. In the 1860s, the Colorado territory was less inhabited and feudalized than Texas. More independent than New Mexico due to having more water, it also held a wealth of lead, silver, and gold mines in its mountains, but the former Texas Republic still dominated with its bigger herds, ranches, and pressing mercantile spirit. It was New Mexican vaqueros, often with slave status, who herded the cattle rounded up by the Anglo-Americans in both Texas and on the then-verdant, grassy prairies to the north. A bitter rivalry grew out of an economic arrogance that somehow, in spite of Santa Fe never being the northern political and trade boundary of its Lone Star Republic, New Mexico remained the Texans' to claim. Resentment of Texas wealth wove quickly and deeply into the culture. Later, during the Civil War, Confederate troops composed mostly of Texans, pushed north through the desert up the Río Grande toward the gold mines. At the bloody Battle of Valverde in 1862, a now famous battle cry rang out, voiced by the commander of the Colorado infantry come to the aid of the New Mexican Union forces: "They are Texans, give them hell!" Over a century later, in the 1980s and 1990s, a defensive arrogance still rode on bumper stickers from the Lone Star State declaring "Don't Mess with Texas" in New Mexico streets. Texans never quite relinquished New Mexico, continuing to run ranches and hold vacation or business property in its dry, beautiful spaces as if they owned it—which they did, sometimes, and do.

The land-linked rivalry with this powerful neighbor has been long, often bitter. But it can't be denied that Texans and the peoples of the Rocky Mountain/Río Grande corridor are cut from a similar cultural cloth. They have had in common a love and knowledge of the land, shared genuine affection for the Indian and Hispanic cultures, as well as valued behaviors of generous hospitality, reciprocity, and cooperation understood in the neighborliness so dear to the Western Spirit. On the ground, the connection with the land, the shared history of ranching and farming, the big sky, the oral history that makes them such entertaining braggarts—"telling jokes on themselves with grandeur," to quote a friend—their hearty enjoyment of natural beauty and scruffy life in the dry land all make some Texans good neighbors. Over the years, many have "immigrated," respecting the simpler way of life, becoming highly regarded environmentalists, reliable board members, or trusted postal

workers in New Mexico. But twenty-first-century survival in the heartland is now subject to pressures beyond the early cattle barons. Real estate is, and big Texas money is, a big part of the problem, its development impacting the many ranches, lower- and middle-income families, and agricultural lands in the West.

It is delicate talking about wealth. As de Tocqueville knew, we Americans are a people who enjoy and identify with money, but great disparity in wealth is at the heart of the land-use dilemmas in the Southwest. At the same time, a good number of resident Texans are just as passionate about the beauty of the Land of Enchantment and the need to protect its resources and cultures as active locals might be. A cultural paradox, carrying hope, all the while accentuating a new class structure that is eroding the geographic and cultural tapestry. Our times are complex.

Before the defeat of the American revolutionary forces in Quebec City in the late seventeen hundreds, Benjamin Franklin had spent some time at the Chateau Ramsey in Montreal attempting through diplomacy to convince Quebec, then Lower Canada, to join the Union and become the fourteenth American colony. Unsuccessful, he left and is quoted in a museum exhibit at the Chateau Ramsey as saying, "It would be easier to buy Canada than conquer it." Such American spirit from that earlier Republic prevails in the Southwest today, as profit continues to drive the colonization of western lands.

In 1926, foreshadowing the late-century's lifestyle, a group of Texas women proposed building a resort on the east side of Santa Fe. The immediate resistance by Santa Feans was successful in putting a stop to the resort plans and simultaneously founding the Old Santa Fe Association. The spontaneous coalition of artists, merchants, lawyers, and businesspeople, mostly Anglo, were, as one of the founders, the well-loved writer Mary Austin, wrote, "promptly found to be possessed of the heresy that maintaining a creative atmosphere is sometimes more important than 'bringing money to town.'"

Speaking of gift economies in his essay "The Ecstasy of Influence: A Plagiarism," Jonathan Lethem clarifies, "A gift establishes a feeling-bond between two people, a gift makes a connection." The resistance to the Texas women's project was the beginning of an official structure to protect an architecture rooted in a way of life. A colleague reminds me that in the association's efforts to inhibit development, the founding members' vision was limited at times by their class's romantic views of the poverty in which many Hispanic people lived. Yes, they were children of their time and culture, progressive and sometimes paternalistic, but here in the 1920s, acting without the impetus to greed, these new-comers from the East or California still knew, cherished, and were committed to the bond that "makes a connection," engendering "a creative atmosphere" rather than mercantile profit.

The gifts held in common in the Southwest are many: the light and sweep of the land itself; from it, the adobe architecture, Pueblo pottery, Navajo jewelry, Spanish and Navajo tapestry; the paintings and sculptures of the Native and Anglo artists. The customs themselves, not only the icons, support the care and cherishing of the gifts: the community cleaning of irrigation ditches called acequias, the Good Friday pilgrimage to the Santuario de Chimayo, where hundreds of Catholics, Buddhist, Jews, and atheists, walk to reach the miraculous dirt that has inspired an agnostic friend to say, "In New Mexico dirt heals everyone!" More well-known are the candlelit neighborhoods on Christmas Eve, the elaborate and restorative Corn and animal dances in the Pueblos—all gifts for ourselves *and* each other as a whole, for the life *between* us. They are the true commons of the culture, as the plazas in the Pueblos still are on dance days and as the plaza at the center of town initially was. These are the kind of cultural symbols that Lewis Hyde spoke of. They carry zoë-life and create the commonwealth among us.

On Christmas Eve, in a bustle of lighthearted activity, friends and neighbors invite you to gather paper bags, sand, and candles and join others similarly equipped at our friend Ranny's house on Canyon Road. You're in the north end of town, which was the original farming and residential area of the city. With festive hearts, together, we set to assembling the farolitos, planting the candles in the sand at the bottom of the bags, then lining them up along the paths, the tops of the adobe walls, and in the older parts of town, along the roads themselves. Lighting these little lanterns at sunset sets off a celebration within a larger celebration. Luminarias, or bonfires, "good fires," are built and lit as well at various corners in the neighborhood. And we know friends are coming. As with the communal care of the acequias in spring, the kinship of neighborly care is present. Winter is a darker time. Rather than for the fresh runoff of spring water, the joy of expectation here is for the warmth of fire, the lilt of voices and, traditionally, lighting the path for the coming of the Christ child.

The spirit is akin to *Natishkuataw*, that the Anishinaabe in the continent's northeast speak of. Natishkuataw, "a place where we come to meet those who are coming to greet us." Akin also to the enlivening spirit one feels while traveling to the dances in the pueblos knowing friends and loved ones from all over the state, even neighboring states, are on the move, winding their way to Zuni Pueblo, perhaps, for December Shalako ceremonies, or to Jemez for a Corn Dance, or Taos or San Juan for the Turtle Dance, meeting together in a beautiful celebration of the land and its people, in a community's prayer.

Here in Santa Fe's neighborhoods on Christmas Eve, they come, too; they gather—friends, family, visitors, tourists—ambling happily, subdued and joyful in the blazing or gentle glow. The streets, still dirt in my memory, are alight from one luminaria to the next where we stop and sing Christmas carols, strangers or loved ones together celebrating the natural joy of larger community. Voices, song, beauty. And hospitality. For tradition has it that homes in the area open their doors to the wanderers, offering hot cider and biscochitos, anise seed cookies—and sometimes a whole buffet—as an antidote against the chill night air. Those who know the custom still welcome us.

In another neighborhood now mostly made up of part-time home owners, a few Santa Feans with a long history in the city continue the tradition of the farolitos, preparing hundreds of bags weeks ahead. Theirs is a quieter neighborhood, less frequented on Christmas Eve than the now highly commercialized Canyon Road area, but the several miles of walks and adobe walls are always warmly lit. In a local living room I hear how the residents deplore that the part-time owners are not pitching in in spite of the tradition being essentially one of neighborly care. One friend's voice wavers as he speaks of relishing years of communal anticipation and effort. "It's . . . practically . . . lonely now, doing so much work with so few," he admits. The absentee owners do come to enjoy the beauty of the Santa Fe

Christmas, but as outsiders, hands idle, watching instead, very unneigh-borly indeed. The celebration along the candlelit paths and sidewalks, this easy, natural event prepared with and for the community and its visitors, seems to have become entertainment, a tourist attraction. A consumer product. A class thing, too. A pretty sideshow not at the heart of their rapport with the place. It was meant to be more. It remains so for some, fortunately, but the insatiate materialism renders the ritual more fragile. As culture erodes, the "song" of the community erodes, their welcome, perhaps, too at times. "In this town," my friend says now, "we weary of being someone else's Third World entertainment."

Sadly, whether Bostonians, Californians, or Chicagoans more re-cently acquainted with the City Different, or Texans linked to its history, many twenty-first-century homeowners have become spectators in the very city they value and inhabit, albeit part-time. In spite of the possibilities of a richer and more complete relationship with New Mexico, they have become part of the pattern of a larger displaced society, while the more discreet culture, breathing with meaning of the land, has been losing ground, its symbols commercialized.

The entertainment phenomenon has led the city to set up police bar-riers blocking off Canyon Road and the Acequia Madre area—the streets marking the original communal farming section of the city—and now the most coveted neighborhood for multimillion-dollar adobe homes. Intended to protect the farolitos and the wanderers from newly increased traffic and behaviors that do not respect the spirit of the occasion, officials are stationed here to question the drivers of cars coming in. Sometimes, though, the intention misses the mark.

Born and raised in the neighborhood where her father built the family home she lives in, a former neighbor of both Indian and Hispanic ancestry found her father obliged one Christmas Eve to go through a checkpoint, having to explain himself to get to the heart of the celebration held at his daughter's home. This grates on her, despite her usually resilient good humor. The courtyard her father built is a place resonant with story and the delight of generations, where three luminarias are assembled and tended every Christmas Eve, where every farolito on the roof edge and adobe wall is placed and lit by hand. Where all are welcome.

Home for the winter in 2008, avoiding the suffocating crush of crowds on Canyon Road, I sought out this courtyard. On this year's walk, it was the only place where the laughter and camaraderie of singing around the fire rang out. Animated by the family's musician nephews and uncles, strangers and friends, part-time and full-time residents, we were happily drawn into the songs in Spanish, in English, for cowboys or lovers, for Christmas. We felt the hearty pleasure, too, as lighthearted jokes on all our distinct cultural quirks peppered the evening. Here was the deep-seated spirit of the place, alive. Leaving this island of care and fire to walk back into the dark, moonlit night, I was reminded how much tourist buses, crowds, and crowd control have meant the loss of this more intimate rapport among the people. I couldn't shake the thought of how insulting barricades are to the very people who built the neighborhood generations ago. They are a fallout, a sign, a perversion of community values in favor of commercial activity.

There were few bonfires elsewhere in the neighborhood—only three that I saw, two strangely untended. At one, a gentleman and I attempted singing but could not distract the others around the fire from their busy conversations about work and politics. At the third, well tended by a stalwart family holding up the tradition, I, along with my fellow wanderers, sipped cider served on the street corner rather than inside the house—

understandably, given the numbers of people, a majority now of strangers. Twenty-five thousand, the local newspaper reported. One tourist, bright eyed and surprised at this ritual, shared her wonder with me. In the streets the scene is still pleasant but, for the most part, social. I amble back toward the Plaza to my car while the crowd thins, the adobe houses, the silver half moon, the candles' glow still carrying beauty, the season's magic. As Christmas morning approaches, the luminarias quietly extinguish themselves toward sunrise when farolitos might be burned for joy.

Electric plastic farolitos abound along the adobe walls of downtown hotels in Santa Fe and cities beyond New Mexico. In this way the soft-lit image of Santa Fe style has spread to brick porches far beyond the Land of Enchantment, without the lighthearted wandering, the fires, or the songs. One could argue that plastic is now the common element of culture, as humble as a paper bag, or that farolitos in either material evoke fond memories of the tradition for those who know it. Even in plastic they can give off a lovely glow. But yikes—plastic! I have mixed feelings. There's that sadness again for another cultural symbol packaged and bought rather than something made together to celebrate our lives and identity. A commodity rather than a symbol of the relationship with the earth-related culture that gave rise to it. I knew now that high-end consumption does not necessarily lead to cultural insight. Neither does widespread consumption.

For a long time the West knew a generous, hospitable wealth that did not mark divides, that simply understood that the spirit of gift protects and nourishes people while maintaining and regenerating their community. Fortunately it continues among some individuals who support projects for training youth in film or agriculture, protecting archaeological sites, or contributing to land restoration. But its spirit is perhaps best summed up by New Mexico's well-loved cowboy-rancher-governor Bruce King, who stated: "You might say that my political philosophy grew out of my upbringing on the homestead, where any traveler in need was welcome, whatever their background. My parents never asked a passerby whether they were Democrat or Republican before watering their stock at our well." A man of integrity instilled by the land and water, he knew the reciprocity of the Western Spirit that is integral as well to the Indian and Hispanic cultures' roots. In the Anglo western culture I grew up in, my parents were familiar with ranchers on the range and miners in towns like Leadville and Central City, whose generous hospitality or financial support saved lives and newcomers' successes before and after the turn of the nineteenth century. Mom would share some of the stories about Horace Tabor's generosity and his beloved Baby Doe's tenacity as she took my sister Kay and me for drives in Denver past the Tabor mansion, a landmark of mining wealth—and a great love story—during the silver boom. In the 1920s, many California intellectuals, like the author Mary Austin, adopted the *Land of Little Rain* and valued its peoples and stories, knowing a community needs the integrity of its land and rituals in order to stay alive and whole.

Once the mining boom came to an end and the railroad was laid down, the Rocky Mountain/Río Grande corridor (as well as California) quickly became a heartland for adventuresome, often rich, sometimes famous easterners at the end of the nineteenth century. Indeed, the first truly settled Anglo-American colony in Northern New Mexico was composed of artists, rooting in Taos as of 1898. As the twentieth century began and wore on, both Taos and Santa Fe drew many of the United States' best intellectuals, writers, and artists from both coasts. The discreet outposts of their lives, where the high, dry climate was good for their health (particularly those suffering from tuberculosis), were also good for their art, giving them the creative freedom of a simpler life, in a quivira, "a quite unknown territory," foreign to the cultural dictates of the upper-class East. Under the radar, these personalities enjoyed a humbler life connected with a land of vast light, starkly stunning natural beauty, and rich traditions among its peoples.

For the Anglo-Americans, New Englanders, Californians, or midwest-erners who came early on, interest in the land and cultures was stronger than interest in their own wealth, stronger than materialism's capacity to erode. Charlie Russell was one of them. The architect John Gaw Meem's contribution to preserving the Territorial and the Pueblo Revival styles of architecture makes him one of the best known. Other newcom-ers in the 1920s were unconventional women in an unconventional land, where reverence for the land was understood and freedom to do good, creative work was paramount. Whether it was Mary Austin supporting the revival of Hispanic crafts, she and Mabel Dodge Luhan campaigning successfully for Pueblo land and water rights, the poet-anthropologist Ruth Benedict collecting myths and stories at Zuni Pueblo, or Laura Gilpin witnessing the dignity of the Navajo in her photography, many Anglo women, some accustomed to privilege, some not, became daughters of the desert, as author Barbara A. Babcock described them. Accepted by the indigenous people, they in turn established respectful, viable, actively cooperative rapport with those whose land they had come to. In town, Allan Houser, Fritz Scholder, and Pablita Velarde—great names in the founding of contemporary Indian Art—were supported by the efforts of

Dorothy Dunn at the studio she founded in 1932. Like Charlie Russell, these women from Anglo culture were drawn by the great spaces and the rich traditions of the Indian and Hispanic people; sharing in them, they also became defenders of their dignity and rights.

That was then. Extreme individualism in the American practice of wealth had not yet set in motion a consumer culture. Now there is an aloofness of the wealthy that one finds more commonly elsewhere, a growing class consciousness, and an innocent ignorance that is often just as damaging. Resilient, a microcosm of forces beyond her own history, the City Different holds her own the best she can, but, colonized again, her soul is at risk. And perhaps so was mine, caught now between the folds of new mantles of greed and the invisibility that poverty creates. I was home and not at home, foreign in my own culture.

FIVE

Tradition: The Art of It

*Art that matters to us—which moves the heart, or revives the soul,
or delights the senses, or offers courage for living—however we choose to
describe the experience—is received as a gift is received.*

—JONATHAN LETHEM, "THE ECSTASY OF INFLUENCE: A PLAGIARISM"

Art and ritual keep the spirit whole, restore and enliven one's affection for the land and a way of life. *Inditas* do this. "Little Indian songs" born of the music, games, camaraderie, and blood exchanged over centuries between the Pueblo and the Hispanics, they reveal the hearty liveliness of the shared root between Hispanic, Indian, and eventually Anglo in the rural habitat of Northern New Mexico. Social or sacred, the diversity of styles in this music of Río Grande del Norte in no way inhibits the joyous "we" as the audience, strangers and newfound friends, joins the musicians in recognition of "Nuevo Mexico, Muy Lindo, Muy Querido," during the annual Nuestra Musica concert. Folk music, ballads of the people. Song.

And dance. In the Pueblo languages (Tiwa, Tewa, Towa, and Keres), the word for song, dance, story, and sometimes prayer is the same. All imply gratitude and celebration, revealing that the tradition of story is enacted by a complex webbing of ritual and memory passed down by and for the people. The winter dances in the pueblos are animal dances—Buffalo, Deer, Eagle, and Turtle—celebrations of our kinship with them. Like the Corn Dance and other summer dances, they are among the twenty or so annual day-long ritual celebrations dating back about nine hundred years.

The Dance is a collective feast with dedicated preparation in the kivas, involving long hours of practice for the dancers who sacrifice much of their time from other obligations. Ritual care is given, too, to the headdresses and clothing—antlers, spruce boughs, wild turkey and eagle feathers, dresses, leggings, kilts, always the turquoise, and moccasins—that, as it is said, must be as beautiful as the earth they walk on. And food. Abundant servings of good food are prepared: green chile stew, posole, chicken salads, turkey, biscochitos, meringue pies, fry bread. Families cook and bake weeks ahead to celebrate with relatives, friends, relatives of friends, and friends of relatives, who will be welcomed into their soon-to-be crowded home. After watching the Dance in the cold of the morning on the plaza, guests bask in the immediate, lighthearted welcome, content to wait their turn to eat at the table. Then served, happily restored by the good offerings, the belly warm, the spirit keen, enlivened by the generosity, we return to the dance.

I met Reina in Chaco Canyon as I migrated home in 1988 before moving into town. Taking time to "touch the earth," I had signed up for a pottery workshop with the Acoma potter Lucy Lewis, her son Drew, and her daughters Emma and Dolores. Five of the participants were professional potters, two of us were painters touching clay for the first time. Reina, also a prizewinning potter, was our cook. It was that twinkle in her eye, her light, warm laughter and goodhearted pleasure at our appetite for her thoroughly irresistible green chile stew that made it easy to be friends.

Years later, her voice carries a cordial excitement, and I am stirred as Reina announces "The Buffalo have come!" At Jemez Pueblo, just as the sun peeks over the mountain on Christmas morning, the Buffalo and Deer dancers appear on the crest of the hill. Each one receives the blessings of the elders as they file down toward the singers and drummers awaiting their arrival in the plaza. At the heart of the village, the brown earth slopes gently toward the elders' shrine and the vivid red foothills behind. Here the "the Buffalo" come on Christmas Day, the day after as well, and on January sixth, the Epiphany, or the Feast of the Kings.

This year, in another crisp white winter, the day after Christmas was quiet as Reina and I, drawn again by the Buffalo, wound our way from Albuquerque to her pueblo at the foot of the Jemez Mountain range. Reina is a friend, an elder, bright eyed, dark haired, and little. We chuckle as we get in the car, saying that, always short, now with age, she may be getting even smaller. But I know the strength of her sweet dignity. It is tall. The comfort of her laughter and the mutual pleasure in each other's presence eases the effort of the hour's drive. Years have passed since my first Buffalo Dance one January sixth, where I was welcomed by Reina and her family. Today as the dry piñon-speckled hills slip by and the sky seems more vast than usual, memories return of their brightly decorated home and her brother greeting me warmheartedly, laughing with delight at my surprise that the Feast of the Kings, "Reyes" in Spanish, was my name day, too. In the Pueblo culture's oral tradition, the day becomes the feast as well of those with like-sounding names: Reina, Rae, Ray. Naturally, I relished my name being part of the glad feast, where well-fed friends related by heart and spirit in the silent breathing of the earth, her winter crust softened and ripened by the feet of the Deer stalking in the Dance, the scurrying of the young Antelope, and the vigor of Buffalo Woman, the Buffalo Men, the run-soar of the Eagles.

On my return a few years later, Reina's brother is gone, but on the village plaza the ringing beat of the buffalo step is as breath-giving as always, and the corn, yellow today rather than blue, warm at the heart of winter, is still held fast in Buffalo Woman's hand as she moves, lithe but strong, with the two impressive Buffalo Men at the center of the dance. Among the Eagle on either side of the Buffalo, there were several Eaglets this year, and the young Antelope and Mountain Goat dancers were as lively as ever, frolicking among the mature men, the Deer, whose antlers wove and bowed in time with their gentle gait.

Summer and winter, for hundreds of years, the Dances have gone on in the nineteen pueblos, "the same every year," an eight-year-old complained to his dad during a Corn Dance at Santa Ana Pueblo. Together with his dad, those of us next to them in the crowd broke out in laughter when the boy had asked, "Can we go, Dad? This dance is exactly the same as last year!" That is, of course, so much the point of ritual, so much the reason to come, to watch, to be there and stay; that it be the same, that it reenact the story of the Dance, that it create again for us a layering of time and worlds to breathe in a timeless state in "exactly the same way." Thank goodness the Corn Dance, the Buffalo Dance, the Turtle Dance, and others bring us the orchestrated step of the dancers, where as humans we can be related by heart and spirit, related in the silent breathing of the earth, our own breath sustained by the dancers' motion, the singers' voices, the heartbeat of the drum. Hopefully, the young boy will eventually be grateful with the rest of us and cherish returning for these days of listening and watching, in tune with his people's rich celebration, offered for the spiritual health of their own communities and those of us outside them. Perhaps he will choose to join the Dance.

Reina and I stand in front of her relatives' house among the quiet, relaxed crowd bordering either side of the plaza. Seated in camp chairs or standing, the gathering of families and friends has its own beauty. Surely it is in the quiet demeanor, the discreet but festive dress, the hair well combed, the wrists and chests adorned with turquoise or silver, and the simple striped elegance of colorful blankets and fringed shawls that ward against this year's cold, contrasting with the "cool," clean, black jeans and baseball caps. But mostly the beauty is in the faces, alert, calm, in tune. The people are watching. Knowing. Alive with the song and the step being offered for us. Familiar.

It is cold today; the Dance will end early, an hour before sunset. At the last Dance, clan members leave the crowd and form a seemingly protective circle around another inner circle that the dancers themselves have formed. There is shimmering of Deer and Antelope feet, their antlers accompanying them. The Eagle dance, erect, exuberant, their wings spread wide before being taken—captured, that is to say—draped over the shoulders of men clan members and carried to the warmth of the home where they had practiced for weeks, their exhausting day of Dance now done. The circle opens while clan members weave back into the crowd; then rhythmic, flourishing, joined in a new configuration, all the animals journey abreast toward us, the Buffalo at the center, followed by the darkly clad singers and bright drums, flanked by the Deer, Antelope, and little Mountain Goats. Earlier, each animal had its own step in the Dance, but here, as the day closes, every foot is lifting, light, vigorous, ringing, in step with the Buffalo. They advance in one compelling movement, one body, one potent song. As they reach our end of the plaza, the music is more subdued but still resonant, while another inclusive circle is formed and this clan's Deer are "taken" home as well.

The gift has been great, unbidden, generous. The resilience of the spirit is renewed. The blankets and camp chairs are folded. Cleansed, the crowd gradually unravels; elders, teenagers, mothers, babies, fathers, and children begin to leave for Albuquerque, for Santa Fe, for Denver, or closer by, like Reina and me, for a relative's home in the Pueblo, to end the day with some of that deliciously hot green chile, posole, green Kool-Aid, Jell-O, the pies, and other traditional fare. Such a feast. Here, as with the Inuit and other indigenous peoples' drumming, art and prayer are one.

Kinship with the land and its inhabitants breathes in both traditional Native and Hispanic art. I have experienced it through my own family, as what we share with the Indian people in our western pride and love of Navajo blankets, jewelry, and Pueblo pottery. The process of cultural cross-fertilization has gone on a long time in the West, from before the borrowing of ancestral Pueblo weaving methods by the Navajo to the use of silver and hornos, "outdoor adobe ovens," learned from the Spanish. Lewis Hyde would recognize the zoë-spirit, the life-giving spirit of these arts grown from the land and the traditions of its peoples.

The Anglo-Americans who came west naturally developed an art as well. Experiencing the impact of the light, the sterling wild, the peoples and their histories begged for expression. The painter Charlie Russell was a wrangler who had arrived in Montana a few years after the battle of Little Big Horn and still rode the range at the time Sitting Bull was murdered in 1890. An inveterate defender of Indian rights, Russell claimed that the incomparable greed of his fellow white man tearing up the earth made the Americans "the new thief" outdoing the most extrav-agant thievery depicted in Native stories. When the railroad brought the end of his cowboy way of life, he set to painting from the sketches of the lives of both cowboys and Indians he had drawn over his dozen years on the range. In them, he depicts some of the rough life one would ex-pect of a wrangler, but also the passionate care for the wild deep in his own core and shared with his Indian and cowboy fellows along the Rocky Mountain corridor as the West was "won"— or, more realistically, stolen by the Americans.

The railroad, steaming its way across the continent at the same time that reservations were being established fostered trading and spawned towns like Gallup, New Mexico, on the edges of the Navajo Nation. A town like so many bordering reservation towns, where Native people have long suffered dislocation from their community. In this case, poverty developed, also very much wrought by exploitation of Indian art, a commercial thievery first established to satisfy easterners' nostalgia for the people who were commonly perceived in the late nineteenth and early twentieth centuries as "The Vanishing Race." Greed is a human fail-ing, and as a cultural phenomenon institutionalized in the commercial mores of a time, it played havoc with a Western Spirit based on cooper-ation and generosity.

Yes, the railroad brought a problematic relationship with Indian art—commercializing it in the late nineteenth century—but there is another tradition, that of the respect and cherishing of Indian art by those from the American culture who identified with the vast beauty of the land and developed close relationships with its peoples. American artists', writers', and intellectuals' support of twentieth-century Native and Spanish Colonial art began in the 1920s and '30s. As a result of the Studio School established by Dorothy Dunn at the Sante Fe Indian School in 1932, painting on canvas was integrated into the Native artists' traditions, joining Pueblo pottery and Navajo weavings and jewelry as art reflecting their own peoples' realities. In 1962, this educational tradition expanded into the Institute of American Indian Art that trains both American and Canadian Native artists, and later to the founding of the Museum of Native Contemporary Arts in the early 1990s. Today's Indian Market, where fine traditional as well as contemporary work is sold, developed from this early collaborative tradition in Santa Fe.

My first trip to New Mexico and my family's life at the mountain cabin were woven into this culture of cherishing. As I grew up, for me and for many westerners, the *commonwealth* of my land, its symbolic feel, its very own sacredness was articulated in the Pueblo and Navajo art more than in the beauty and adventure in paintings by Charlie Russell, grown out of a more individual art and masculine experience.

It was my Irish mother who introduced me to Charlie Russell's art, knowing the spirit of adventure and power of the land that we both cherished could be found there. It was she who, in a small shop in Grand Lake, first pronounced the word *Acoma* to me, as she showed me an exquisite piece of the black-and-white pottery from that ancient and living pueblo village perched on a tall mesa in the southern New Mexico plains; it was she who shared, almost timidly, her experience of the Sun Dance she and my father had attended in the late 1930s.

It was she in Santa Fe that year of the ten-year-old, who bought me a thunderbird ring from a man selling jewelry, as the Native artists still do, under the *portal*, the porch, of the Palace of the Governors. A thunderbird! How I loved that ring. I can still see the long narrow silver body and the oval turquoise in the middle. The thunderbird, the eagle, oh, but our eagle this time, a western image, not an American USA icon. The meaning of the world was still western, wrought from my land.

In my home in Quebec, far in time and place from the dry Colorado and Río Grande plateaus, I still have the picture of the three Navajo in Canyon de Chelly that hung in its rough frame at the family cabin. And warm and present in my studio there hangs the nineteenth-century Navajo blanket, quite tattered after seventy-five years on the floor in the neighboring bachelor cabin, but cleaned, washed, and dried in the Colorado sun, then packed and carried to Quebec. We carry what the land gives us.

In traditional Native art, a quiet alertness, a contemplative relationship with the natural world can breathe. As with the ceremonial dancers, the artists are aware of kinship and a responsibility toward the community. Aware of the rare tradition they are entrusted with, different pueblos have periodically closed their Dances to visitors from outside the community in order to protect their meaning from being treated as a mere show. We can ask, early in this twenty-first century, given the new way of the land, what has happened to the symbolic power of Navajo and Pueblo art for their communities and the larger Southwest culture? The paradoxical appropriation of Native art as the mark of belonging to the commercialized Santa Fe style has had a dangerous edge, threatening to appropriate the symbolic expression of the local culture. The strength and vision of the peoples, though—and the deep and jubilant collective work at survival through ceremony, art, and literature—lend beauty, hope, and sustenance to the authentic contemporary tradition.

Human beings need symbolic meaning. Do these symbols of Southwest life and tradition still "circulate among us," as Lewis Hyde suggested? Yes. At the Dances, the wearing of jewelry by the families, the elders, the singers and drummers, the dancers themselves bears witness to the power and spirit of the tradition. This art is still integral to the spiritual meaning of the culture. Creative exchange still goes on among the families and artists, with communities in Mexico as in the early times, and with other of today's Indian nations in a vibrant trade, spanning the continent and crossing into Canada.

"The way we treat a thing can change its nature," Jonathan Lethem asserts. And so it is that in the American culture, to a certain degree, the traditional icons of Navajo jewelry, weavings, and Pueblo pottery have undergone a conversion from being collective symbols to private symbols reflecting the privileged purchasing power of the few. Still cherished by many, however, who may not be able to afford them. There is paradox and complexity in this cultural area that is reminiscent of the complexities in our relationship with the land and its development.

In the real estate takeover, it is worth noting that Native people living on the reservations have been protected to a great extent, as the reservation land, belonging to the federal government, was and is not available as private property to the newcomers. However, like during the first American colonization, the relationship to their art has been very much a part of the late-twentieth- and early-twenty-first-century changes. One can ask: Is the purchase of Native art a repetition of the old theft of the railroad days or something new? The Native people themselves offer more elaborate discussion of this, but it is clear that sales of contemporary Southwest Indian art have contributed to a healthy prosperity for a good number of families and communities, allowing for the continuance of ceremony and hospitality.

SIX

Resources

What I fear most is despair for the world and us:
forever less of beauty, silence, open air,
gratitude, unbidden happiness, affection, unegotistical desire
—WENDELL BERRY, FROM GIVEN, SABBATHS 1998, IX

The sun was bright in the parking lot of the Motor Vehicle Department. I stood there stock-still, new driver's license in hand, suddenly seized with the realization of how much Americans were left on their own, left especially to their own budgets. New Mexico had called, I had come to the wild land I loved, but it hit home that day. Like other retail workers, archaeologists, business employees, and artists I knew, I would not be able to afford car insurance or, even less, health insurance. Keeping a roof over my head was a preoccupation. But it was not just New Mexico. I was stunned realizing the US society as a whole held no common agreement to give basic economic support to its citizens. My shock at this discovery was surely augmented by the fact that I had been a resident of Quebec for twenty years, where the Canadian consensus for taking care of social needs was still operative. The understanding there was that, as a society, we have a common responsibility to take care of certain basic conditions, after which people can fend for themselves. Basics like health care, unemployment, affordable car insurance. In Quebec, housing too— whether renting or buying—remained very affordable to people with a middle-class income. Not so now in Santa Fe, where a simpler life used to reign. Working people experienced a gap between the rich and the poor more familiar to Central and South America than to the bounty most people think of as the United States. The millionaires here had their own kind of pressure. I laughed with a wealthy architect friend from Chicago when he joked about his predicament, "How do you make a small fortune in Santa Fe?" I searched for a reply. "Come with a big one."

The mountain woman I was had become a resource, like other educated and talented men and women equally exploited who, as one construction worker friend pointed out, "get off the bus in Santa Fe every day." Like them and like many New Mexicans who had always remained in their home state, I found no reliable living wage. After five years, even with my numerous jobs rent money was still not predictable from one month to the next. It was hard to shake that dull-to-the-bone sense of irrelevance that poverty gives. I kept busy, stepping in ragged time to the Santa Fe shuffle, lucky that my knowledge of the land and natural affinity with it—along with my professional training—served the needs of the expanding tourism industry.

Glad to be absorbed in fulfilling my various contracts, I was in my element as a guide leading visitors from eastern museums, Canada, Poland, California, or North African government committees through the tales the adobe architecture told of history in the New Mexican capital; accompanying them to the Dances at Acoma, Jemez, and other pueblos; introducing them to well-known artists in their studios; camping with them in the limpid black of Chaco Canyon's night; or interpreting for French-speaking visitors from the State Department. I was eager, as one client put it in her thank-you note, "to weave together for them the history and the spiritual in the land." They said they would never forget; I haven't forgotten either. I was as exhilarated as they were discovering their strengths—and balance—climbing narrow trails onto mesa tops, and as deeply moved when their new awareness of Southwest peoples opened their hearts to fresh understanding.

Being a guide in southwest art and archaeology had the advantage of a high-paying job in the Land of Enchantment ($10 per hour since the budgets came from out of state) but the disadvantage of being seasonal—full tilt all summer, dwindling contracts in the fall, and none in the winter. I related as Victor di Suvero, a seasoned Santa Fe poet, had the audience chuckling at a recent poetry jam when he recounted the familiar questions he often receives about his career in poetry: "Why do you do it?" the curious might ask and, always, the clincher: "Does it pay?" We know in North America poetry doesn't "pay," nor does art, although it's a little better, and southwest archaeology is not too different from art. But these worlds have the great advantage of creating community. In my jobs, my heart and mind flourished in warmly reciprocal relationships with guiding clients, museum and archaeology personnel, fellow writers, and conference attendees.

Planning and coordinating stimulating literary conferences for a local nonprofit and organizing poetry readings at an independent bookstore introduced me to some of America's favorite poets and writers: W. S. Merwin, Galway Kinnell, Gerald Stern, Terry Tempest Williams, and Native America's acclaimed authors N. Scott Momaday, Luci Tapahonso, and Linda Hogan. Their work, so crucial for giving voice to Native people and their West, carried into contemporary literature the light, movement, and shelter of the land as I had known it and still felt it.

Archaeology too allowed me to work closely with fellow-minded people and kept me out camping on the land. Early on, a new friend, Nancy, Navajo/Cherokee herself, invited me to be an educational consultant on rock art for Utah's White Mesa Institute on their field excursions in Canyon de Chelly. There, Nancy's friend and his son led us through the canyon, sharing Diné stories their ancestors had handed down and pointing out a specific rock plateau, its red sandstone starkly beautiful against the sapphire sky. I listened, engrossed, standing this time at the foot of the place where the Navajo had taken refuge from Kit Carson and the US Army's assault. The land's memories continue to thread through our lives. At the end of the day of hiking, observing, and exchanging stories and good food, we settled comfortably for the night on this ground of kinship, beauty, and history, in the restful silence of a campsite next to the family's sprouting cornfield.

Coordinating conferences in town also meant meaningful relationships with curators at the Museum of New Mexico or the School of Advanced Research. There was the privilege as well of assisting in collaborations between Hispanic or Native artists and the curators when the artists came to study the stories invested in their peoples' art and artifacts protected in the collections. Working closely with Native and non-Native anthropologists, archaeologists, artists, and museum personnel, who not only knew the tenacity one needs to survive in the desert but also revered its treasures, I reveled in grounding my knowledge more thoroughly. New Mexico had called. I had come home.

In my first years as a guide, I felt I was both witnessing and sharing our living culture with the visitors. Eventually, however, aware of the changes impacting the local traditions, my conviction in telling the stories waned. Until so recently a land of its own, its own *patria*, the Southwest had not developed a conforming culture. That was its strength and beauty. But now, the sparsely settled populations were being made economic outsiders, and I with them. This is the rub: being made irrelevant to one's place. It is economics again. But the paradox of cultural

survival continued; the richly diverse communities resisted being assim-
ilated. And I, still willing to hope I could stay, energized by the passionate
concern with this land we had in common, went about life reveling in
the good-natured spirit among us.

Naturally, I nourished my art in the field when I could. The pottery
workshop in Chaco Canyon where I met Reina was important for my artist
life. Lucy Lewis, fragile in her old age, was made comfortable in the workshop tent by her daughters, Emma and Dolores, as she accompanied our
discoveries. Gathering the clay, grinding pigment, chewing yucca stems
to make our brushes—perfect for different widths of line—boiling Rocky
Mountain Beeplant to bind the clay. Childlike delight had us, the Anglo
participants, dancing about, slapping each other's backs with clay hand-
prints while we rolled out long sheets of it on the canyon floor to dry. Later
we sat, the silence between exchanges almost breathless as we shaped
our pots and polished them with river stones. The next step was painting
them, an even more breathless task, but first they needed to dry.

"How long?"

"Until it's ready." (Emma and Dolores were very patient with our
questions about time.)

"Where?"

"You see the little niche in the cliff?"

"Yes." And there I placed my first piece of pottery, light now, in the
sun, in the niche in the cliff where its clay had come from. Tears sprung
at the emotion. Earth to hand to earth, to sun, to earth. *Beauty above me,
beauty behind me, beauty all around me.*

Back in the studio, after resisting for a while—"I'm a painter, I have
to paint," I told myself—I would give in to the joy and long-lasting sustenance of sculpting with clay as my main medium, using smoke-firing

and raku as firing methods. Seeing the new work, Isabelle, a sculptor and studio neighbor, suggested I might be interested in being part of Nicasio and Janet Romero's annual outdoor sculpture show near the village of Ribera, below the mesas east of the capital. Supportive and welcoming, Nicasio and Janet and their community were rooted, lively, and open, and I was moved by my first experience of "land art" among them, placing my sculptures amidst the dry stones of a fountain in a side courtyard. I had offered to read some poems for the opening events that traditionally brought together musicians and a well-loved flamenco dancer. On Nicasio's suggestion, his musician friend Carl Bernstein accompanied my reading, offering a moving experience which began a new friendship and a long-term relationship of music performed with my words.

As often as "the shuffle" allowed, I headed for the tranquility of my studio in a communal adobe enclave beside the Santa Fe River. With the ebb and flow of the water to accompany me and the collegial calm of other artists working in their studios, I could feel my life was almost complete, as I shaped and polished clay sculptures, printed, painted, or drew. Here, an occasional commission and my line of graphic designs for posters and T-shirts added to the trickle of income arriving from so many directions. Among the well-chosen things packed in the Mazda that had brought me south were my records—yes, vinyl LPs. Classical music, jazz, French and English folk songs, bluegrass, Johnny Cash, everything to fit the varying moods of living. Music always accompanied my work in that happy place, an antidote to the lonely discord in my survival activities. In the hours settled there, I tapped into the simple joy of my roots and wrote, reflecting more mindfully on my circumstances.

> "America the beautiful" still, yes, for starlit and spacious skies. I love my America, but prosperity has fallen into the hands of so few. I worry about the impoverishment of our culture in spirit matters, about the void, the lack of sustenance for its men and women. The new image says money is it, is all. If money is all there is, then it determines individual human value—so when money is scarce, our very existence is threatened! I object. We the people are so much more.

Often on a summer evening I could seek out friends' company in a shared garden just above the river at a studio friend's home. Together, five of us had each planted our own section with flowers, herbs, or vegetables and, in a ritualistic spirit, built extravagant scarecrows. After bringing water up from a frequently recalcitrant old pump in the river, my fellow gardener Ranny and I might end the day sitting by our seedlings, watching the plants grow in their mounds, small irrigation ditches surrounding each one. These were not drought years for the garden; the bright gentleness of afternoon rains could revive the weariest of souls.

Such a good life I had come to, except for that struggle with debilitating poverty and the shock at the erosion of my beloved culture of the wild! My dad had been right. Though I had always known of New Mexico's simpler, more rural life and had looked forward to it, I was too late. Another world had been grafted onto it, and the material underpinnings of my own life were lost in the seams.

The challenges of scrappy survival wore thin. The intellectual, cultural, spiritual wealth in me held, but the physical resource I was ran dry under the strain. Not ill yet, but unconsoled at the deeper shock of the cultural and economic shifts around me, I wrote to my sister Mary. Freshly home in Colorado from many years working in Egypt, she had commissioned a painting to mark my return.

Dear Mary,

I'm here in my studio thinking of your commission for my version of a "kinder, gentler America" painting. Well, my dear sister, the truth is, I have yet to experience the States as a kinder, gentler place. The real truth is, I feel it as devastatingly materialistic, terribly self-indulgent, expectant of privilege, impoverished in spirit, aggressive and violent in attitude and deed. This land of Americans does not feel like a safe place, but a land of mindless greed. I have experienced a great sense of loss of the sacred and my own tenderness for the world trying to live in this daily "marketplace" culture.

That said, the earth-land of desert sky and mountain, Indian country, and Hispanic warmth that breathes steadfastly and discreetly by, here, amid the larger society's gross and hopeless impositions, revital- izes my heart and sustains a great number of strong, kinder, gentler individuals who care deeply about people and the vulnerable gracious beauty of Northern New Mexico: its acequias and canyons, the Río Grande and prairie grasses, the deeply woven and wildly eroded desert life, its eagles and ravens and rain, and the clay of its arroyos that branch through the land. This is what I love, identify with, this is what brought me home. This is what is being terribly colonized. It seems Americans are just as vehement in their colonization of their own as they are of others. It's the speed and aggressiveness of it that gets me, "kills me," I almost said, but I won't let it do that. I had so wanted to believe that this was a cohesive society, that fragile hope you spoke of, but I see no evidence for it, as the exposed beauty and

meaning of what's here is grabbed up and packaged with no thought of the value, the life worth of the cultures and land that could, if respected, help sustain us and give us all ways of finding better direction for a much-changed world.

This phenomenon is much exacerbated here in Santa Fe, and at first I wondered: is this the States, or is this just Santa Fe? Now I'm realizing this is really what's happened to the States; it's just more visible because there was never much of a middle class-buffer here in the first place.

In the rooted kindness, the simple solidarity, the heartfelt generosity of people here, I recognize a truly western American openness that feels natural and good, and is what I needed as my life changed. I also came home for the "can do" American attitude, as opposed to a formalism and "can we do?" of the Quebec mentality. Unfortunately, there is here a travesty of the true pioneering spirit; it has become "Me, can buy!" Unfortunately again, I can't compete with million-dollar buyers, so we'll see how else I can manage to live close to what I love and do my work. It is great to be able to write to you, someone who knows that cultural alienation is a reality and not just a gripe. And your respect is doubly precious to me here. I feel very lucky to have art to respond to life with, otherwise I would die with all this travesty of meaning about.

Love to you and Seena

In the refuge of the studio, the strains of Jacques Brel's song, a litany of human losses, moved me: our failures and death at the end, our unfaithful lovers, poisoned birds, disheveled cities. He sings, *"Il n'y a plus d'Amérique,*" "There is no more America," the mythical land for many world peoples, for the individual, the middle class, the land of promise, of fulfillment, of unbridled "pursuit" of all our dreams and desires. I feared it was true. Our wars revealed this. Our wealth had us floundering. Our conformity stagnated our culture. E pluribus unum implied a coexistence. The ten-year-old I was had believed in the peaceful coexistence already present in her western life. Only much later did she learn that in the United States, among "many" of the early colonies, women, Indian, and black people were excluded from the possibility of belonging. Her generation worked with civil rights, women's rights, much of the unfinished business of what had been the Melting Pot ideal. A full generation later, in the solitary peace of polishing a sculpture, she mused on how the compensatory value of gratification, above all, perniciously envelops us: at the grocery store, in the sexual hype of the media, even in some churches or spiritual movements. In her early forties, from her own homesickness, she knew how the absence of a sustained relationship to the land saps us of the courage to test our solitudes, and sabotages our efforts to delve into our inner springs that carry the promise of a different way of being. Her thesaurus gave synonyms for "to live" as "breathe" and "walk the earth." This was her simple cherished ambition. Looking back, I feel a tenderness for her dark-haired beauty, her intense awareness, her pain. Even then, before sub-prime mortgages, her morning began with the blank pages of her diary and often with concerns for her culture:

> Under the graftings of greed bereft of ideal, consumer conformity has not been enough to ensure our values or give us a cohesive, consistently peaceful and fair society.

Us? We the people? From the vantage point of New Mexico's "un-American" cultures (Indians, Hispanics, and Anglos who like living with Indians and Hispanics), the Melting Pot, in its drive to create a successful, mostly white, one-dimensional society, could not tolerate deep difference as an integral part of its thrust. Its singular drive to forge a prosperous middle class did not recognize the engrained thoughts that culture is.

What about the wealth of our languages, our accents, the lilt and ring of voices that we vibrate to? And the Celtic harp, the eagle bone whistle, the biscochito, the Irish potato, the drums for the Buffalo feast, the mariachi band, the polka, the bluegrass, the dancing! Tango, the French wine, the evening song as focus for comfort and celebration? Treasured ways running deeply in our genes weaving together multiple facets of identity, making life livable. Somehow we have dislocated their rich diversity in favor of holding conformity at the center of identity. Such a paradox.

Cisco slept on the worn but cozy green couch, both of them appeased by Keith Jarrett's *Köln Concert*. A fresh sculpture, the clay still damp, lay on her worktable as two friends knocked and wandered in for a break. While the sculpture dried, they laughed and mused over it, intrigued with "She Ain't Dead Yet," that had started with the idea of an angel figure but had become, as sculptures often do, something else. This one was now an aging woman with legs kicking, strong, and a haunting but funny hand transforming from human to bird claw. They fell silent a moment wondering how the sculpture might be connected to their friend Isabelle's depression and threatened suicide, but did not dwell on it. They had come to touch base about her plight, and together the women agreed who would accompany Isabelle and when, so she would not be alone.

Before leaving, they shared their fellow feeling about the country, an ongoing conversation. Having come to New Mexico in the 1960s during the "hippie migration," an influx of adventuresome Anglos of her generation wanting to go "back to the land," these two had found means to stay on and make a simple living. But this did not protect them from the pressures of the Melting Pot, increasing in the nonconforming land they lived in, changing the parameters of what mattered and of the local art market. Consoled by the exchange, not alone now in her struggle, she read to them from her diary: "For sure, melting is not what we need now! Enough is blended away anyway in the uniform world culture with our phones and jeans. Above all humanly, hard as it is for the hundreds of millions of us, we must grasp, understand, accept that the beliefs and customs of others, what they hold dear, are actual realities, cherished realities."

Tanya, upbeat now, reminded them she'd heard, "The melting pot has melted, but the people haven't."

"Yeah," Dan continued, "so here we are, no pot, just land and people!"

Laughing good-heartedly, they quipped, "People are so wily about surviving and thriving."

"We won't melt!"

"It's a New Mexico thing."

"Or maybe it's an artist thing. We're tough, no fat, just sinew and bone!"

Cheerfully now, the visitors moved on with their day, she and her little dog with theirs, walking up past the studio buildings to the river, her supple gait contrasting with the worried thoughts that hung on about her people. Soon seated at a worn table under the cottonwood, her pens scratching accompanied the sound of water.

A media consensus has supported our American acceptance of consumption as the very purpose of existence. This is the menacing value of our tumbling, trembling culture. It is failing us. It is failing our ideal of e pluribus unum. Are we failing ourselves?

We must begin again, she wrote, together or alone, steadily, simply, deliberately; and continue with that sense of miracle and pleasure, of hope and joy in the full sense of human possibility that we know in our knowledge of the unseen, and the unsaid, to be real.

As I closed my dairy, Cisco leapt up and bounded happily ahead of me toward the trail. Delighted, I savored the late afternoon where, among the cottonwoods and chamisa, the sage plain, and aspen forest, I could still visit parts of the balanced, more peaceful wild I was fortunate to know. Here, for now, I lived, walked, worked hard with, and enjoyed the company of people who felt the resonance of the word *land*. Would it really last in its own changing rhythms until the mountains turn to dust?

II. THE GIFT

THE QUIET SPACE WITHIN
Ceramic sculpture in Ribera acequia

SEVEN

Grand Lake

Why did I come home again? It was true, as my dad felt, my life was so full in the North. Fearing the struggle with poverty he was sure awaited me in Santa Fe, he couldn't understand why I would leave Quebec. A successful teaching job, close friends, enjoyable colleagues, a people with a real identity and a culture of song, where somehow I was welcomed to belong almost like family. Yes. But it was not enough. The same hardworking, worried father had often said, "Without the mountains we don't know where we are." This identity with the land was strong. Lest we forget, at this writing, places to experience and really know the wilderness are already rare. Access to it in our highly urbanized and bureaucratic world is even rarer.

Growing up in a big family I could feel crowded in our city routines. In what we called "Grand Lake"—not the village but the neighboring wilderness seven miles outside of the small mountain town—there was room for each of my parents, my six sisters, and me to experience solitary moments in the woods, or to ramble along the deer trails together, enlivened by our curiosity and wonder. There was mischief, too, hiding cigarettes for my big sisters under that huge boulder, "the smoking rock."

And then the privilege, rare for the younger children, of riding the seven miles with Mom to bring horses up from the town stables to the cabin.

That Grand Lake welcomes me now, acres of barely touched beauty. The cabin itself breathes quietly with the myriad gestures of thoughtful, happy people paying simple, unobtrusive, joyful attention to it and each other. Alone today, the memory of each familiar gesture—chopping wood, lighting the stove, opening or drawing the curtain—renews the gentle air of kindness. As I arrive, my cells wash through themselves and shuck off the tensions and weight of the world. My body "smiles" at the sudden naturalness of passing from being to doing, and I get to thinking about the chemistry of beauty, a deep repose; a cellular sense of being part of the woods and clearing. Integral to the chemistry, too, is the jeweled timberline depth of Blue Lake, a three-thousand-foot hike above the cabin. A revelation of crystal water, the freezing cold of the glacier spring feeding it, trickling down through fat bouquets of purple columbine.

The familiar, rusted, hardly legible Bowen/Baker "No Trespassing" sign hangs loose on a board near the porch, and I pray: "Forgive us our trespasses as we forgive those who trespass against us." Why does this come to me, this notion of trespassing? And forgiveness? Perhaps I feel the Ute's presence in this, that their old summer hunting ground still needs to be healed. Perhaps my cherishing it can matter there, too.

Rarely has anyone seriously trespassed here in our time. Only once was the chemistry of beauty altered to distress when I arrived with my sister Kay one August. The shock was devastating. An angry hiker? A druggie? Someone possessed had vandalized the cabin, breaking windows, kicking in the door. Probably he was only angrier still upon finding extremely faded furniture, cherished forty-year-old calendars, links to older relatives dead and gone, nothing valuable for the drug market or other consumer markets. No phones or stereos either, for there is no electricity.

Do I forgive the man who violated the cabin? Beyond the shock, I feel nothing personal. In a way I see him as the first sign, an omen, an entity that revealed the beginning of the end. Individuals perhaps can be forgiven, but the society that produced the violence needs to be throttled! While Kay recovered on the bridge over our beloved Bowen River, and I set to getting help for repairs, the destructive and invasive act sank in and revealed the new fragility of our tacit sense of commons in that cherished land. I felt, more than I thought, at the time, that this violence was part of a larger imbalance in our contemporary society which the cabin wilderness had afforded us protection from. It remained our sanctuary for a few more years.

Fortunately, the invader didn't touch the faded, well-loved print of three Navajo in its old hand-hewn frame: one man with two women in velvet skirts riding horseback in Canyon de Chelly. Like the Navajo, my mother was striking with her black hair and dancer's erect posture, although her hair was from a different place, "Black Irish," she said. This daughter of my adventuresome, fisherman granddad had taught her daughters to ride early on at our paternal uncle's ranch outside of Denver. At four, I remember my wonder at seeing the larger world from my perch on a buckskin mare named Patience. Later at ten, twelve, fourteen, this picture of the Navajo gave me my first image of the beauty of womanhood, speaking to me more directly than Denver's newspaper images of elegant country-club women. Womanhood on horseback. As we did in our mountains, the Navajo, the Diné people, walked and rode in Canyon de Chelly, free and safe among the animals and arroyos of their desert canyons. They knew the beauty of their land—and of women who

rode free. These women, like my mother, were models. Confident, graceful, and free in the wild; here, at any rate, there was no contradiction. We
were women of the West.

The Cabin

I'd like to tell you more about the place. Let me tell you this:

In this West, Space is the actual place where earth and sky join
whole
where the ground of experience is not only the dry or
flowering clay beneath your feet
but is, too, the great silence of air, breath, your breath   light
and the absolute activity of sky

In this West, when we said "this land" it meant all this
this space, actual place, ground of loyalty
this West, this land, vast
not a void   but a bond
embraceable by breath
and attention
to elk in the meadow, paintbrush bloom on the trail,
the flight and stillness in its cliffs
our greetings to the sun.

I must let you know what's happening to it.
I want to tell you why it matters.

This is a real place, a family place, the root.

"The cabin" is really two log cabins. What we called the "bachelor" cabin had come first, built, family history says, in this forest in the 1840s by my dad's relatives, the Clarks, as the fishing club of their company, Cocks and Clark's Printing. More rats winter in the shadier bachelor cabin but leave us the comfort of its worn Navajo rugs and its generous open hearth for the summer. Not far, across the stream into the clearing, just past the sparse remains of Gaskill, a short-lived gold mining town of the late eighteen hundreds, is the Bamber cabin built by Great-Uncle Clark's friends, Mr. and Mrs. Bamber, under the Homesteading Act in the 1920s. Their home, sturdy on a few stone pylons, allows the mice a more separate territory. Two quiet shelters, standing with aspen and lodgepole pine where, agreeing to their need and ours, the animals leave us the woodstoves and tattered couches for the summer. The Bamber cabin, the couple's daily home, the clearing, the aspen and lodgepole pine wilderness it is a part of have been my home: two hundred acres bought and paid for by my parents and three other families who shared it with us, when, after Mr. Bamber's death, Mrs. Bamber wanted to leave. A gift, this home, already consecrated by all those who had gone before when it sprung gently into my childhood.

Let me tell you how to get there.

First, with your dad, you've got to pack the old, green Woody station wagon, loading all the food and bedding on top so as to leave room inside for your mom and six sisters, too. Then you drive north from Denver, along the foothills of the Rocky Mountains that call you and that you will later understand are the spine of the continent. Then you turn west through, say, Big Thompson Canyon (there is more than one way to go) and start up, and up, until you reach timberline on Trail Ridge Road. There the tundra blooms, vigorous and fragile, contrasting with the immense, craggy, foreboding peaks across the steep glacier valley. Ahead are more peaks, but mossier looking and rolling away blue to the horizon. If you follow them, you can see all the way to the Medicine Bow in Wyoming. On those clear days, and they were actually all clear days then in Colorado, you did see forever.

Now, drop gently down through lodgepole pine forest into a long meadow valley, sparkling—as green, I think, as Ireland—flowering with the tenacious beauty of wild penstemon, paintbrush, and sometimes, if there's enough rain, the tiniest wild orchids, each alive in their short alpine season. Here you leave the highway, turning in where I've left my red bandanna waving on a branch for you. Drive slowly back through the willows, the peaks of the Never Summer Range glowing above you. Cross over the Colorado River we are so grateful to, because it starts here to make the continent's watershed, the Continental Divide. Go on past that one lonely corral where you sang "Home on the Range" once with the cowboys, when you were very little. Where the Baker river winds, you stop to look for trout and see the beaver's work for this year. Meander. Linger. In the sun. Move on, now, past the elk meadow. Watch for them if it's dusk. Look for that tree where the bark is scratched up by the porcupine; here, enter the clearing.

As usual, the cabin's chinking is a little loose, but the place stands firm (even the outhouse weathers this year's winter well). Yes, please do. Roam. Alive. Alert. In joy. If it's cool, someone stokes up a fire in the great old woodstove. You relax into the very, very worn furniture that has the ageless comfort of our family songs around the fire and our sound sleep in the black and crystal nights. Peruse the poem on the wall beckoning you to the Jeep Trail, reminding you that *you are of the fabric of the*

wilderness, this beauty is also your own.
Moments you know of this land whisper back to you:

I crossed timberline, and wept for the beauty of the tundra, forever

I, and the end of day, hush
on the stillness of high green-forested rock
still with the willow
run through with silent rapid breeze
spreading swiftly in the glow of coral ice

When I am still, when the doe recognizes me, stays with me, it is real; when
I speak in silence to her as, alert and seeing, we bond in recognition, I know
she is my life-kind, though not my species

In this place you'll find soft, clear water that you bring up in buckets from the Bowen river. You'll find space, both towering and comforting, and pine cover, dry and moist. You'll find Dad teaching you about the stars at the tip of the pines at night, about the berries to be found and the miners' history of scattered logs in the underbrush. You'll learn the exhilaration and strength of who you are by riding and caring for borrowed horses with your mom. You'll open the door in the morning, early, to the sun, and its light will dispel the night's crisp chill.

There are no fences, so sometimes you'll find hikers who have lost their way. You give them water, what they need, share the sightings of the moose in the valley. Make sure they're safe on their way.

It is a holy place, a place where you, your family and friends, the neighbors scattered every mile or two in the valley know, share quietly, love what it is to belong to the intimacy of the wild. The Bambers knew this, as did the Ute hunting groups for hundreds of years before them. You cherish this knowledge and, in your flesh, care for these meadows and forests with the two, three, now four generations that mingle here with your own short life.

EIGHT

The Sacred, A Letter

Sunday, January 23, 1994

*Deer Dance day at San Ildefonso Pueblo with songs and drums,
cornmeal, soft prayers, and the dancers coming over the rise with the
sun while here in Montreal, snow falls in the cold city air.*

*Dear Dad, dear Mom, dear sisters Mary, Gonie, Peg, Jo, Kay, and Anne,
I am glad that, in spite of limited leisure in my Santa Fe life, having
lived closer to our home territory for five years makes you all feel even closer to
me now that I'm here again in the North.*

*I'm sending you all a copy of the letter I received from the superintendent
at Rocky Mountain National Park responding to my letter requesting a few
more years on our lease. The response seems clear—a clear "no." It seems
there was a special directive from the James Watts Department of Interior in
Washington, DC, about eight years ago, telling the parks they must stop extending inholder "reservations" (what some of us have called our "life-lease"),
as they had been doing since 1906 in accordance with law and custom. The
directive is the main reason for the "no." However, the west side superintendent is going to take a last look into it to see if there could still be the least
"mechanism" to renew the reservation legally. We should receive a copy of
the directive and info on the plans, if any, they have for the cabins themselves.
I had understood that the cabins could be taken down and the land returned
to wilderness, used for backcountry rangers, or dismantled and resold at a
national park auction. According to his understanding, since it is not desig-*

nated an official wilderness area, people would still have access to the area on foot or on horseback.

This recent news from Washington's "eminent domain" really means I need to turn my efforts from trying to extend our time at the cabin to letting go of Grand Lake as we have known it. As I begin, I need very much to be in touch and share with all of you. I know some of you have already started letting go but, I think partly because of the shock and pain of discovering so many "used to be's" in what I loved so much in the West, I couldn't really start, can just barely start now.

I do know that next to the gift of life, Mom and Dad, your gift of the cabin being our summer home, that wilderness being our place to roam, to be alive, to be alert and in joy, has been the most important gift in my life for me. Indeed, the two gifts feel very much like one in the measure of my consciousness of my own life and self. I am grateful, and grateful too for knowing that we share the knowledge of how special, how sacred a place it is for us.

You may remember going to a reading by N. Scott Momaday in Santa Fe when you visited me last June. This fall, coming across an article of his on sacred places, I was flooded with recognition and tears, reading:

> Sacred ground is in some way earned. It is consecrated, made holy with offerings—song and ceremony, joy and sorrow, the dedication of the mind and heart, offerings of life and death.

"Holy with offerings." This is so true at our cabin. Your gestures of thoughtfulness, like so many offerings, inhabit the place: the love that came to me finding my jeans warmed by the fire Mom and Dad had built before we woke; the times that we sisters have done something similar for each other, or for the newer children; the wildflower in the cup with Chris's photo that you left me last summer, Kay. And so young, little, close to the ground, all the myriad moments that brought me to enter the world—at age six, was it? To see, to know it as beauty, wonder, and reverence. Playing in its mystery, basking in awareness of life much greater than I, but peaceful and serene. I entered it on those days of collecting berries with Dad, especially collecting blueberries, so magic because they were blue, and on those days of horseback riding with Mom, feeling so strong and exhilarated. The being together, all of us around and about. Sometimes when the big girls would allow, hiking together up the Jeep Trail to Blue Lake and leaving the orange peels behind so we could find our way home, or going with the big girls into town! Hiking to the beaver ponds, earnestly stomping out Uncle Bill's cigarette butts as I went, cleaning the fish by the cabin with Dad, or down on the bridge fishing, watching sunsets, putting worms on the hooks for the nieces and nephews when I was older because their own dear dad was squeamish about it. And little again, seeing Indian paintbrush for the first time, my face flushed close to it, struck with heated joy at the wonder of that red plumage and its name.

Then, of course, our evening ritual, walking to the meadow to see the elk, or the adventure some nights of getting to go for a ride in the car in our pajamas, keenly, as you'd say, Mom, peeling our eyes for the deer. Peering through the beauty of the night to know again their fleeting or stillness. Their exquisite grace and shyness quickened me then and soothes me now. Somehow in the doe's leaping beauty I recognized the core of my own unique aliveness and knew the quiet truth of my own shyness. I was such a little girl then.

"Song and ceremony." How many times we sang from that green and white songbook around the fire! How many times far from there those songs have comforted me, even here below the vast shimmer of the northern lights in Quebec. How many times, Dad, you revealed the stars to us at the edge of the reach of those tall, straight lodgepole pine. I easily conjure up the black beauty of the sky and the comfort of your presence as I, in pajamas, listened, *learned too, but, being so involved with the vastness of the beauty, had a hard time always getting the picture of "the Bear" just right!*

"Joy." I know the simplicity of ecstasy for being at the cabin. Joy. Joy and sorrow. Yes, once, seeing your parents' hearts torn, my sister's anger, and me feeling powerless to help. Joy as an adult, a new flowering, when Mary, Seena, and I spoke the first lines of the poem "The Jeep Trail" while we sat with Kay near the lightning tree and buttercups, after a long hike through silver logs and breath-giving surprises of columbine delighting us on the trail.

"The dedication of the mind and heart" in so many things. The cooking and chopping wood; the hauling of water from the stream to drink and re-fresh, to soothe and bathe in; the sweeping, cleaning, the packing; all the caring among us; the returning. Carol and George's attention to it skiing in on cold winter days. Then my solitary, homesteading summer of painting and writing, with its new realizations of spiritual purpose and dedication. Yes.

"Offerings of life and death." The bats Kay and I so carefully buried, the screams of the deer when the mountain lion attacked, the celebration of the fish we caught. Mr. Lasasso's stopping by with his generous offering of fish from his overample catch when I was homesteading alone at the cabin.

> To encounter the sacred is to be alive to the deepest center of human existence. Sacred places are the truest definitions of the earth. They stand for the earth immediately and forever; they are its flags and shields. If you would know the earth for what it really is, learn it through sacred places. . . . There the earth lies in eternity.

From that first day, I feel the cabin was given to us somehow already con-secrated; already so fine, pristine with harmony and wonder; so very alive with the light, my cells felt, feel, all one with the space, the river, the meadow, and my family. I'm sure the Bambers' care for it, and that of the Ute during their summer camps before them, made it so for us and those we have shared it with.

To me it is a bright and gentle ground for our lives, sprung into being in a complete kind of joy. It has brought much learning that no other place in this world has allowed me; the dear familiarity of the land we belong to, my knowing of mystery and eternity, and the simplicity of human love. It has been truly home and heaven on earth to me. Sometime soon, it looks like we will have to give it away or back, in some way, to the mystery, to itself. In fact, to the "eminent domain" of the government in Washington, DC. I don't quite know what that will mean yet for my spiritual balance, for our family, for the cabin itself.

Language and the sacred are indivisible. The earth and all its appearances and expressions exist in names and stories and prayers and spells.

The sacred music for me goes like this: Colorado, Longs Peak, Columbine, Taos, Arapaho, forest, the Jeep Trail, Blue Lake, Grand Lake, the Bowen and Baker Club (that's us!), blue spruce, Río Grande, Trail Ridge Road, Medicine Bow, the mountains, Kawaneechee, tundra, Mesa Verde, Santa Fe, canyons, cliffs, the meadow, elk, Canyon de Chelly, Chaco, deer and rainbow trout, German Brown, and the first note, the origin note, Grand Lake/the cabin. It's a long song and will ring fuller and truer for me as it mingles with your memories and thoughts as we move toward our good gathering—four generations deep! I keep thinking about what we will do for a ceremony to say goodbye to this place. A bonfire, songs, poems, stories? I look forward to how it will evolve.

I must close now. Does anyone want to hike to Blue Lake this summer?

NINE

Eminent Domain

I must tell you this:

You cannot stay, you cannot return.

Thirty-two years ago, your father, hearing that Rocky Mountain National Park was annexing land, thought it wise as a long-term protective measure for the cabin area to have it included in the annexation. In his sixties, he no longer relished the upkeep of the two wooden bridges that we crossed and the cleaning out of the beaver dams beneath one of them. Negotiations were begun, ending in three of the families selling their shares. Ours gave the park our land and the cabins in exchange for use of them until 1995, at which time the "reservation" for our "life use and occupancy," written now in the name of my father's seven daughters, would automatically be renewed.

At that time, there was a long-standing tradition of a trusted spirit of land protection among locals, rangers, and the Rocky Mountain National Park, part of a tradition and a culture founded with the parks themselves in 1902. Reservations were automatically and legally extended as family members grew up and continued to cherish this land, wanting to remain. This principle of humans inhabiting a national park is well understood in Europe, in England in particular, where many national

parks are inhabited, with farms and villages found within their boundaries. In the United States, before the problematic legacy of James Watts,
the Reagan era's Secretary of the Interior responsible for both the Bureau
of Land Management and the national parks, in-holders (people like ourselves with these reservations) and park personnel together protected
the valley's streams and animals from poachers and welcomed the
strangers who often visited. Now this long-term cooperative caring and
weaving of human belonging to the land will no longer exist. Based on
a 1988 special directive from Washington, DC, no extensions can be
negotiated, and all the land has become solely government property. Our
family had to leave in 1995. The remaining local community was gone
by 2001. Only one single, fortunate, aging inhabitant's efforts at negoti-
ation were met with recognition by the government. She was granted
permission of life use and occupancy of her cabin until her death. Our
family's disinheritance, this community's disinheritance, is typical of the
centralization that generates greater loss and dilemmas in the West. It
is reminiscent too, a friend says, though in a much milder form, of the
larger tragedy Native Americans have suffered. Yes, in a much milder
form. In spite of the laudable vision Theodore Roosevelt had in establishing the national parks to protect wildlands for mainstream America,
many Native and Hispanic communities were dispossessed of their land
at the time. The Havasupai of the Grand Canyon, for example, were "re-
located" to make room for "people" in his new national parks, according
to Roosevelt's own explanation.

In both cases, there is grave danger for the human spirit. At the root
of this centralization I see the insensitivity and fear of a puritanical mind
toward the earth and those whose lives are led in relationship with it. It
has become, here, excruciatingly apparent, and risks taking control. As

in many parks, in Rocky Mountain National Park two populations remain: campers, fortunately, and government personnel, no longer on
horseback, often, ironically, infesting the forest with rules rather than
welcome—an aberration of the tradition our culture held. It is worrisome
seeing law enforcement personnel roaming in my family's and neighbors'
places. Worrisome too that very few interpretive rangers will be educating
hikers on our trails. Devoted volunteers who sometimes replace them
are rarely trained with the appropriate knowledge. With the budgetary
demands of our failing military structure in the States, the parks are becoming part of a military property relationship with land. The argument
of using natural resources for national security established the purpose
of the national forests in the States, but the national parks were meant
to protect the land and animals themselves, independent of such logic.
There is serious cause for concern. Government policies have reduced
funding to the parks and oriented the monies elsewhere, undermining
both the rangers' work and the visitors' experience. Coupled with the
increase in visitors every year, the dominance of regulations—rather
than the protective and educational mission on which the parks were
founded—risks generating more crime control (and crime), so that
eroded trust and erratic violence rule the day.

I would not want to think this possible, but I have already seen it at
the Grand Canyon and, later, at Chaco Canyon, where three gun-sporting
rangers accosted a small, respectful, enthusiastic group of five visitors
I was guiding, not once but three times. Suspicious, following us, reap-
pearing at our every change of focus on the trail, they harassed us until
I spontaneously delivered a heated lecture to them on respect for the
public trail, my group of visitors, and the sacredness of the petroglyphs.
We were left then to go on more happily with our visit.

Indeed, standing policies and current pressures from Washington for oil and gas development on public lands raise the question of whether the American people could lose the parks altogether. In 1995, the Senate Resource Committee's spring draft bill that proposed abolishment of the national parks in the name of economic necessity was stayed. Over fifteen years later, though, the Arctic National Wild Life Refuge, home to the Grant's caribou and the Gwich'in people, is still seriously at risk of being exploited for oil in the name of national security. Again. At all costs. In a society impoverished by its incapacity to think in other than quantifiable, materialistic terms, the argument of national security justifies the argument of government control over place—the mutually cherished commons of this land and its people. A national greed. A greed we can see that has fed our wars abroad.

How does the West go on? In New Mexico, the Southwest Regional National Park Office and its archaeological lab where I did lithic identification in the 1990s, is nearly extinct. There remains now the National Park Service Intermountain Regional Office based in Denver, Colorado, an administration that covers a territory from Montana to New Mexico. The Denver Intermountain office still has power, but there are many fewer, too few, regional offices for adequate on-site care of the wide land and water resources.

For the Southwest, the problem is acute because 90 percent of the resources are archaeological—very different resources from the majority of national parks north of Mesa Verde. Available money and power are not only diminished for the western parks but for archaeology in particular—the resource that offers us the most insight into the human way of life in the wild and continuity with living peoples today. As a result, in Santa Fe plans were made to move the remaining sixty people on the

staff of the Southwest Regional National Park Office into the Old Santa
Fe Trail Building, a historic structure, well-loved for the beauty of its
Pueblo Revival architecture but small, with inadequate space and, very
importantly, no archaeology laboratory! To add to the demise of archaeology in the parks, retiring archaeologists are, in accordance with recent
policy, not being replaced. From twenty archaeologists in this Santa Fe
office, there will be two. The parks, where for a century Americans have
come to know their land, are struggling to survive.

In the Kawuneeche Valley near the Bowen and Baker Gulch that shaped
the clearing for our cabin, the West Side Ranger and Management Offices
of Rocky Mountain National Park have been closed. Overseeing the West
Side streams and valleys and their wildlife as well as visitors is to be done
from the other side of the Trail Ridge Road Pass, from the East Side
offices of the park, where an administration is centered in the highly
urbanized town of Estes Park. For now, campers will continue to have
access to the Kawuneeche Valley on the wilder West Side area and will be
able to experience the pristine forest and glacier lakes. Still, for the
National Park Service, campers constitute a transient, easily codified
population that will continue to be told how long they can stay and what
they can do. Necessarily, for some, I know, perhaps a majority, because
they have not had a chance to learn nature's ways in their own communities. I hope they will be able to continue to discover, gently, respectfully,
this valley, but the drive-through, bumper-to-bumper traffic on the road
to Grand Lake does not augur well for human culture or for the animals.

The loss of a local community leaves a major gap. Gratefully, there
are a few individual volunteers who sometimes are knowledgeable
enough to try to bridge it. Within the national parks on their own land,
the Havasupai, the Sioux, the Diné, and other Native communities now

often educate visitors through the stories of their people's particular loss, personifying their capacity to survive and often thrive in two worlds. But in the Kawuneeche there will be little witnessing. Gone are the people whose insights over time have been finely tuned to the complexities of this region's biotic community. Too few will be the people themselves, carriers of regeneration, steeped in the wisdom of knowing how to care for the land and animals and, especially, how to let them well enough alone. Here, reciprocity with the earth was not just an idea. This shared culture afforded a congruent, coherent sense of what it means to live in nature. Time, changing seasons, changing patterns of animal migration, and intergenerational relationships are what made these wilds a common ground, literally, for a truly human society, for its individuals' as well as its communities' identities. Without the actual place and its life, identity is weakened, much less knowledgeable, uprooted from a shared base of reality.

Theodore Roosevelt has been quoted as saying that the people's presence in and enjoyment of the wild was relevant to the safety of the nation. A statement by the founder of National Parks that to Native ears might ring with a terrible irony. But relationship with the land is important to all peoples. Still, one can ask, what poverty of imagination now leads a people who were once capable of inventing a free country to grind it into one of bureaucratic control, creating a country no longer, by far, "for the people"?

The forest was still safe for us during the last years of our time at the cabin when I homesteaded there alone. The mountain lion and the bear never disturbed us, having enough of their own hunting ground. Growing up, we respected the animals' territories and they ours. Today, with the urbanization of the wild, a new population has entered the mountains and the foothills. In some mountain towns joggers and, sometimes, mothers letting their children play do not always know that, at dusk, the trail at the edge of the meadow and the forest is hunting ground. In exurban foothills, unsuspecting suburban dwellers sometimes leave food out for the deer, drawing the tawny creatures and their predator, the mountain lion, even closer in.

In the mid-1990s, with increased human intervention, the biotic balance was already changing. The clear, cool water of our rivers began to carry the danger of giardia, a difficult kind of amoeba. We could still haul water up to the cabin, but not drink it fresh. The last day of my homesteading, I found bear paw marks on my car in the morning, at the trunk, on the driver's door handles, and the front windshield. I had been to the grocery store in town the day before. Obviously this was a bear familiar with human cars and food. A different West. Giardia, socialized bears, and increased crime in the Grand Canyon.

Barred from reentry by our gate, now changed into a national park gate, we were told to surrender our keys. The family gathered that summer of 1995 at the Estes Park YMCA camp, traveling during our reunion over Trail Ridge Road to spend a day with the cabin—a day to let go, a day to love it again, together. I don't know how many of us there were, too many to count, but four generations of us were there. A spanking new baby among us. We sang and wept for joy. We would try to let be. After our gathering around the fire pit, my nephew, David, came to me with a little paper boat he had made from the Jeep Trail poem and suggested we sail it down the Bowen river. We did, from that shore where, as a child, I had cleaned my first fish. We had to let go, each at our own rhythm. Perhaps I am still trying. My life on this planet has been fundamentally altered. The psychic family bond woven from this place that carries our memory and renewed experience as part of the various life forms in the mountains of Colorado has lost the concrete nurturance of its root. As an individual, I have lost a place where my spiritual and physical balance was continually restored. Our loss of the family's reservation is a part of a dilemma in the relationship with land in the West, where loved land in one's own habitat is often government owned.

The palpable sorrow at the loss of places like this in the West is not that there is no more frontier but that there is no more habitat. The "great loneliness of spirit" that Chief Sealth (Seattle) spoke of over one hundred years ago is rapidly compounding in this new century. Barry Lopez points out, in *Arctic Dreams*, that to understand an animal separate from its environment is impossible; the animal and the environment "require each other." I agree. Complex animals that we are, human embodied souls need not only to protect from afar but also to be in actual touch with their sacred places, like the fragile tundra flowers whose beauty and vigor is sustained by the high-altitude conditions. A body needs the earth. It is a question of habitat. We have known this. We have denied this. May we come, quickly, to accept it, to act on it. May we believe.

III. KINSHIP IN EMPTY LAND

COMMUNITY
Ink, acrylic, and found objects

TEN

Empty Space, A Bond

*If geographic place rapidly changes in a way that demeans its natural
integrity then children's early attachment to land is at risk.*

—RICHARD LOUV, LAST CHILD IN THE WOODS,
SAVING OUR CHILDREN FROM NATURE-DEFICIT DISORDER

Coming down from the mountains over Colorado's Berthoud Pass, you are not alone in feeling the palpable loss of your family cabin to the eminent domain of the National Park Service. Musing quietly, you wonder how it will feel to live without that habitat. You stop the car, take in the tapestry of stars with your sisters, the Milky Way a firmament of its own in the black silence. Carry on now the peace of this interlude, wind down on the road again into the foothills until you catch your first glimpse of the blaze of Denver's lights spread out on the eastern plain. A beautiful sight. Consoling. You know the neighborhood you'll rest in is among the lights, about four city blocks around, with graceful cottonwoods that stand guard on your corner. Together you and your sisters remember that it was the wealth of the cattle industry, aviation, and of course, having enough water in the "Great American Wasteland" that made your town a prosperous one. Life was good here, simple.

In this city we could walk the three and a half blocks to and from school, often crossing through the alley and vacant lot on Detroit Street. At the end of the day, green uniforms were shed for jeans and boots, better outfits for playing outside. With a friend or two, my sister Kay and I head for the long, roomy stand of pine trees at the end of Mrs. Simms's garden. A hideout, of sorts, where no adults come; a shelter for our imaginations. Bordered by a tall hedge on the other, the alley side, it is private, an almost dark place of manageable horizon where we invent adventures, campsites, and stories, delaying the return to family at the supper table for as long as possible, pretending our watches aren't working, waiting for the sound of our mother's voice calling or Dad's whistle.

Out in the open, near the front of Mrs. Simms's yard on Milwaukee Street, is a grove of lilacs and other small trees where we play the games that need a fuller expanse of ground for western history to come alive. In our own versions of hide-and-seek, Sitting Bull's friend Annie Oakley shoots game among the houses and trees from here; Calamity Jane escapes to this bush to ward off ambush (saving herself and others with her); and a crazed Baby Doe Tabor fends off the do-gooders who would take her away from the Matchless mine and land she loved and claimed as her own. We mostly know Mrs. Simms through her yard. Not a close friend of our family, but a good neighbor, she leaves us to the joy of our own world within hers knowing, as our mothers did, that freedom means actual physical freedom.

The vacant lot in the next block is part of our territory too, a range of our own. Joining our friend Anne, who also loves horses, we become those proud animals, sometimes wild, sometimes ranch stock, galloping and whinnying among the brambles and dry plants, our scarves tucked into our jeans for tails, flying in the wind. Other times we're jumpers from the eastern equestrian tradition, leaping over obstacles of logs and branches that we can adjust to different heights. Further down on Seventh Avenue Parkway, there are the big ponderosa pine trees, our most regular retreat for climbing. The parkway offers trails, too, for less arduous activity, and abundant summer flowers in the sun.

Our territory here is more limited than at the cabin; there are fences and streets, some of them busy. But we are still free. At ten, our bikes give us a wider range, blocks and blocks of exhilarating rides, stopping almost a mile away for tea at Mrs. Madden's, another good neighbor. The land and our imaginations are uninterrupted.

Three horses need exercising south of town, not far, at Anne's grandmother's farm: two jumpers and a five-gaited Welsh pony with the most comfortable back I've ever ridden. Independent teenagers now, we trot and canter, splash bareback through the Highline Canal, the main irrigation ditch, and across the fields that have not been sprawled upon. On an overnight, we crawl through the barbed wire to catch the horses and delight in a moonlit ride. A good ride, down to the horse's sweat, healing a barbed-wire cut on my leg. We are safe. The only occasional danger is drivers who don't always know that honking at us makes the horses unruly.

Back at our house on Fillmore Street, few cars go by. On the small, grassy hill in the front yard we play King on the Mountain, make snow angels in winter with our cousins, collect caterpillars in the cottonwood trees, revel in Easter egg hunts, and the rain; run out barefoot in early June to hop off the curb and wade in the warm creek flowing at the corner on Eighth Avenue. The backyard is the place where we tasted both bugs and the nectar from honeysuckle; where the swings swung and the teeter-totter creaked; where Mom cut our hair in the sun; where, as teenagers, we washed the green Woody station wagon, finally ours; where, on visits home as an adult, I took such pleasure in trimming Mom's roses. This was where—small, sitting in the freshness of the grass, alive with solitary joy—I first embraced the universe, the light, the immense sky; or it embraced me, all one with it. Whole. Years later, I would learn the words cosmos and sacred, still aware that in the distance the mountains glowed, due West.

On days when they beckon, the family might drive out of the city toward them, across a wide expanse of plain sloping gently up toward the Rockies' Front Range. So soon after leaving the house, the wide-open space is meandering outside the car window. The timeless, half-hour drive carries the familiar delights of our arid land's beauty: the uncountable hues in its rocks and cliffs; its mesas, proud with loaded history; tufts of sage and rabbit brush; the stalwart yucca, whose flowers explode seed into the sun; and the surprise of the discreet ranches it harbors. The few cars on the road can hardly be called traffic, so there's room too in the empty space for that faraway United States as we count license plates: "One from Massachusetts; that one's from Arizona; and, Mom, where is Arkansas?" We count horses: "I got two pintos, five palomino, three roans!" Playful, rising with that tranquil plain as we drive, we bask in the sense of the holy wilderness we know from higher altitudes. Excitement mounts as we approach the foothills and sites we can name: the hogback formation and red rocks, the much-loved, fifty-million-year-old natural amphitheater with its fossils intact.

Just into the foothills we turn, coming to the town of Morrison and Uncle Dud's ranch, where we play with our cousins in the haystack, learn to ride horses, giggle over the squealy piglets in the barn, see cows milked. Not allowed to roam the upper mesa where the cattle are, Kay and I with our cousin Ralph often opt for collecting tadpoles in the river. The long rise of the Front Range is all of a piece, part of the gift of our land, connected to the yard, the vacant lot, the trees, these "wild places" of our town that are part and parcel of the air, the sun, the mountains that meet the dry plain. Changing into our jeans and those liberating cowboy boots (how I loved my boots!) meant all this.

It's here, in the expanse to the west of town, that the first new dominion of sprawl would later shock and dismay me as I visited home, as did the pollution that grew from the hugely increased number of cars per capita. The starkly etched beauty of the mountains against the horizon has lost its glow. Most days, a yellow haze sits and wafts and, in winter, on days with thermal inversion, the brown cloud hangs, obscuring the sky, the view, and some people's breathing. John Denver was right, "It's enough to make a grown man cry," a grown woman too.

One summer, stopping in Winter Park, a mountain ski village, on the way back to Denver from the cabin, my sister Mary and I were enthused at finding a motto there: "Mountain Women Are Natural Resources." We reveled in this. Mountain women: hardy, hearty, self-reliant, resourceful, adventuresome, beautiful with the vigor of the land they identify with, rightful proud. Later the expression wore thin, a disconcerting reminder of how a resource is perceived in our culture.

If land was "empty" in the Colorado Front Range, what does it matter if its pastures and ranges, slopes and dirt trails are developed, built up and paved over? What does it matter if losing the smells and sense of adventure and discovery is killing to men's, women's, and children's hearts? It matters because the amazing density of development means the Front Range is no longer common land but a wide spread of plots owned and parceled into separate units. This empty space, this open country was part of the gift we knew as a culture. Held by either the county or private owners, it was ours, too. A breadth of land where we belonged.

Distinguishing, in *Harper's Magazine*, between commodity and a public commons in a gift economy, Jonathan Lethem asserts, "A commons belongs to everyone and no one, and its use is controlled only by common consent, like the silence in a movie theatre [it is] a transitory commons, impossibly fragile, treasured by those who crave it, and constructed as a mutual gift by those who compose it." Lethem also makes a point about language that can be made for our wide open spaces—"That a language is a commons doesn't mean that the community owns it; rather it belongs between people, possessed by no one, not even by a society as a whole."

This wide land, "between" the city and foothills is lost now to common consent, common enjoyment. Like the edge of a meadow, the Colorado Front Range—edge of plain and mountain—is clearly a major place of growth, driven by the force of real estate and consumer retail profit. Beyond its filled-in suburbscape, the land slopes upward, still dry, somewhat empty, and beautiful. Beckoning. Exurban development, now spottily scattered in the foothills themselves, will continue to encroach, albeit more slowly due to some public open spaces and private lands that are unavailable for development, thanks to ranchers and farmers setting up conservation easements or simply being unwilling to sell no matter what the price. The sprawl. What else to call it? We all know now what it is. It's still spreading. We all lament it whether or not we live there. The sky will hold, but not the air; the land is not holding so well. Do we know what sprawl means besides temporarily cheaper land or a bigger, more elegant house with a mountaintop view? The news is not getting any better. "Western Futures," a report from the University of Colorado's

Center of the American West, offers maps showing housing density along the Colorado Front Range in the years 1960 and 2000, and projected density for the year 2040. If the development, the sprawl, already seems overwhelmingly invasive, we best brace ourselves for the future. The 2040 map shows a widening swath of urban corridor that has doubled in size; the area along the foothills is completely filled in by our urban, exurban, and suburban development from the south in Colorado Springs through Denver, up to and through Loveland, then beyond Fort Collins in the north. Our empty space, our open country, has definitely been claimed as a "somewhere," bought and built, or rather a patchwork of individual "somewheres," in the culture's materialistic terms. Individual ownership over individual, family, or collective stewardship is the norm now. Where are the people coming from? Where is the water coming from? Where will our food come from? Where do the children play? "Empty?" "Nowhere?" This perception of "nowhere," of land unidentified for industrial society's use as a void, has occurred again and again, turning the ground of being, of identity, of cultural and family story, of food and survival into money, fragmenting the land and, with it, our capacity to know, wonder at, and be restored physically, psychologically, and spiritually by nature. We need more than the shopping centers in common.

Cabins in alpine forests, meadows, vacant lots, wooded garden edges, lilac bush groves, trails in the national parks—these places we love or have loved individually and as a people—these natural relationships are also the wellsprings of a healthy, responsible society, nomadic or settled. In Richard Louv's words in *Last Child in the Woods* I find hope for recreating the kind of whole we experienced in that Denver childhood. In arguing for connecting open spaces for our children in San Diego and our now-fragmented cities, he asserts, "To achieve this, however, the public must see the currently isolated canyons (or in other cities, other disconnected natural areas) as something larger and singular. For that to happen, the biological, educational, psychological, and spiritual value of open space must be clear."

Shall we see this soon? In fact, the spaces rolling out to the Front Range are gone, filled in with houses, hi-tech buildings, oh, and how many more shopping centers! Driving west from Denver, we still thrill at approaching the mountains, but one can only wonder at why—how is it!—our society feels "deserving," as the ads would have it, of oversized homes and multiple garages that are so out of proportion with our needs, undermining the needs of the earth for us and for future generations? In losing the physical and symbolic space to sprawl, it is undeniable that the groundwater is dangerously diminished, that the reliable food sources of pastures and rangeland are shrinking; undeniable too that we lose a real freedom that allows us the possibility of knowing the land as sacred and the wholeness of our lives in it. Growing up roaming in a world all of piece was not a restlessness, but an exploration and an entering into being alive with the natural world. Unrelenting in this twenty-first century, with the restless accumulation of oversized private properties, we remain the "feverish children" of the nineteenth century that Walt Whitman spoke of in his poem "Passage to India." In 1869, the great American poet is feverish himself with admiration, "a new worship," he calls it, for the "voyagers, explorers / . . . engineers! . . . architects, machinists . . . !" who had succeeded in opening the Suez Canal and laying a railroad across America. Yet he admonishes them to create "not for trade or transportation only, / But in God's name, and for thy sake, O soul."

Later, poignantly, he asks, "Ah, who shall soothe these feverish children?" We might respond poignantly too, knowing that we are now so tuned to cars, manufactured wealth, and the computer that we have far too few people to "speak the secret of impassive Earth / . . . [to] bind it to us"; too few to wonder with him, "What is this Earth, to our affections?" As a society, we hardly know anymore. Or have we only forgotten that we are made to move and breathe with an awareness of what the waters, mountains, deserts, and their inhabitants mean to us? On a mountain drive, I am reminded.

Species

There are always signs, down the road from Aspen,
STEEP GRADE, FALLING ROCK
and DEER leaping in boundaried yellow space

until seen mangled
fresh, faun blood running, on a fine, stilled hoof
dragged to the side
like the putrid skunk, belly bulged red,
or the drab, broken porcupine
on the roadways
and kills

I know well the deer, still, close to dusk,
taut, pungent, heat swollen in the air
serene, shy among the willows
browsing the meadows fresh keen
fleeting
along the roads

where I saw the fellows of the slain Raven
gather and grieve
offer twigs and alert silence
in the middle of the roadways

where one, late, light gray night,
a man dropped, soundlessly, as in a dream,
gracefully, off

his motorcycle
dead
drunk?

dead
on the roadways
among the kills

The natural world is our habitat as well as the porcupine's or the deer's. We are among those species that need a whole habitat if we are to stay healthy, not overwhelm ourselves with noise, be able to drink clean water, breathe fresh air, eat safe foods, and especially, if we are to be at peace.

On the ground mobility has given way to a culture of wealth and voluntary displacement on a large scale. In conditions like these, how can our culture carry us back to the life-sustaining rapport with nature we need? The West is now the place of development. Habitat is more than views. It is not about square footage. Seeing forever, though, or at least as far as Medicine Bow in Wyoming from Trail Ridge Road in Colorado, was essential to the knowing of wonder and natural, cultural treasure that brought us together, making us what we are. The way the light reverberates in your own core matters. You are imprinted with the chemistry of beauty, the chemistry of knowledge of the rivers and animals that run in that blue distance, of their ways and your own rhythms. Of your safety and theirs. You know what survival in the wild means; you know what intimacy with the wild is. We know together. The chemistry of memory, its smells and tastes of wild strawberries, of meadow grasses, sustains you. You need to be there, to breathe it in, to move in it. We are place- and planet-specific beings. We pretend otherwise. We still want to believe, and the ads of course make sure we still want to believe, that our greed, the outrageous consumption of land and man-ufactured goods, is the pursuit of happiness. But the earth is divided and measured out now; only the privileged few have a place to roam. There is not enough water for the pursuit. Less and less room. Yes, numbed to this, busy, like me, getting somewhere in our cars, we do adjust—but at great cost. More than a family cabin is lost.

Habitat means more than a two- or ten-acre subdivision lot. It is the larger ground, literally, that we stand on to survive and prosper. It is the place our bodies and imaginations inhabited. It stands for us, for this West, for a life we lived and can still cherish. How we shape the habitat reveals what we have come to mean. Empty indeed. If property values are all we hold dear, we are truly an impoverished people.

ELEVEN

Querencia

If we want to be happy at all, I think, we have to acknowledge that the circumstances which encourage us in our love of this existence are essential.

—WILLIAM KITTREDGE, A HOLE IN THE SKY

When I took refuge in Quebec's north after the five years of New Mexico poverty, it was my sister Peg in Santa Fe who would send a bowl of *la tierra bendita*, "sacred dust," over the border to Montreal, to ease the shock of that city's cement and asphalt on my soul and aching feet. It was she who would phone to tell me the meadowlarks had arrived, or that a Rocky Mountain bluebird had flown across her path.

One November morning, having come home to the Southwest for the winter, I step out of her house to greet the sun. The property is on a hill in a semirural neighborhood just beyond the busyness of the old New Mexican capital. Behind my back on the other, northern slope, the suddenness of the sprawl, out of bounds now, chokes the land and the wild peace with the "little boxes" that Arlo Guthrie sang about. In the name of affordable housing, condos mushroom without land or character.

In the quiet of this morning, though, it is the moment that counts. Facing south, you can still feel the breath of the wild in the wide spaces of air beyond, along the slope of the dusty hill, along the chill on your skin, and the freshness of your own breath. A sparsely inhabited land is there still carrying the solace of the untamed, as the prairie rolls gently up and away from me, flattening to meet the Ortiz Mountains further on. The neighboring farmhouses and trailer homes sit quietly, scattered. The breeze crinkles through the juniper trees, and across the land the light of day overcomes the pink of sunrise as the moon sets below the curving horizon. In Santa Fe, very much so in Taos, sometimes in Denver, this breath can still be felt in the everyday air, a commons between us, the people, the birds, the elk. A reminder that we are alive when linked with the land.

This quality of wordless joy, a *querencia*, with the mountains, desert, water, and air is surely what de Tocqueville meant by *those delicious solitudes of the New World* as, drawn by the smoke of a campfire, he readied his boat to join an Indian family on an island. These "solitudes" are the mark of kinship with the land and its inhabitants. They are known in the silence of the land westerners share. The silence, the solitude are synonymous with the wild, the "deliciousness" of the sacred. It is an essential part of the kinship openly celebrated in ceremonial Pueblo Dances. It is present in the community care of the acequia.

Las acequias are, humbly, the "beds for the water," a network of narrow irrigation ditches that thread through the arid land in New Mexico, an engineering feat and ineradicable cultural reality. Allowing for the control and transfer of water, the ditches initially arrived as an integral part of the communal and family land grants awarded by the King of Spain in the seventeenth and eighteenth centuries, and by Mexico later in the early nineteenth century. Embedded in the mountains and valleys of Northern New Mexico, identifying bloodlines and boundaries, the land grants are handed down from one generation to the next. Like the

French-speaking presence in the northeastern region of our continent, the Spanish presence in the Southwest has been lasting and uninterrupted, still deeply rooted today.

A descendant of one of the original land-grant families in Embudo, Estevan Arellano knows firsthand how, for over four hundred years in America, *las acequias'* traditions have sustained the earth and the bloodlines of the Hispanic families, shaping the land and its history as farming and sheep grazing developed. Greeting us warmly in Spanish, the historian, farmer, and translator Arellano continues to delight and move his Albuquerque audience in English. Born of the belonging his own history gives, he generously shares both his traditional knowledge and his scholarship, reminding us that the "noble" Moorish eleven-thousand-year-old (amazing, but accurate!) irrigation system crossed over the ocean to arrive in Northern New Mexico with the already mestizo Spanish cultures in the sixteen hundreds. Estevan's story, vigorously truthful but humorous too, about our historical and contemporary divisions over land and water sounds the depths of his love of *las acequias* and their people, stirring the wide and deep *querencia* in our own hearts.

Querencia, the ancient word, the scholar tells us comes originally from hunting vocabulary, referring to the place animals return to "to spend time in, to eat or sleep." Its usage for humans, he points out, also identifies our "inclination to return to a site" where soul and soil are connected, where the affection or longing we might have for a favorite place of reciprocity and responsibility toward the land make it home, give it anchor. People with heart knowledge— evolved from a close acquaintance with the caves and cliffs, meadows and pastures, watersheds and forests—carry it on. The word is closely akin in meaning to the French *terroir*, which speaks of a place where land and soul create identity and a response that bonds community.

The bond in the Southwest is with the water, of course, that dire necessity, but it is also with the mountain and the natural timing of snow melt from its heights that promises survival again in a harsh agricultural land. This is what the community acquiesces to, responds to, in the spring—a reciprocity known to farmers, herders, and ranchers around the world.

In a valley outside of Dixon, meandering through the orchard together after years of not seeing each other, I bask in my friend Sylvia's easy gentleness and her equally kind welcome. The Rio Embudo tumbles full, close by, snowy peaks gleam in the sun, while nearby Jesse, Ernesto, and Dylan, the men of the family, with three young workers, reshape the banks and gate of this acequia that will feed the apricot and apple trees already in bloom on the Atencio family land. On this day, moving among these trees, the warm earth underfoot, I breathe in again the naturalness that the long-lived rapport between land and people gives.

Throughout Northern New Mexico, in the towns of Taos, Dixon, Santa Fe, and elsewhere in the heights and plains, villages and farms of the dry land, the acequias are a communal resource, a shared responsibility. Each spring, before the runoff from the mountain flows, the *mayordomo* of the ditch puts out the call for the acequia cleaning, a necessary and cherished tradition. Land owners, or *parciantes*, neighbors, interested workers, and willing children gather and greet each other and are soon laying hand to rake, hand to tree root, trash, and stone, combing the ditches clear of winter debris. After a few days of labor, the countryside is adorned anew with fine ribbons of softened earth running among the lightly greening plants. It is as if the land has been, yes, caressed, as well it has, and is breathing better for it.

The parciantes, water rights' holders, members of the Acequia Association, each have a headgate for the irrigation on their land and submit to the mayordomo's judgment about whose gate may be opened first, at what time, and on what day they will receive the acequia water. Once freed, by the force of gravity the welcome snow melt feeds each rectangular field that slopes—sometimes imperceptibly—toward the nearby river where any unused water returns. A seasonal, necessary time when this caring for the land and water supersedes tensions among neighbors.

Born and raised parciantes Eleanor Bové and her husband Phil, the Acequia Madre's *comisionado*, or elected commissioner, are the key people who carry the querencia in my former neighborhood, passing it on not only to their children and Santa Fe's children but, in the generous spirit of welcome of the West, to the newcomers who are willing to respect it. It is they who gather and greet us, make sure everyone has their donuts, shovels, rakes, and gloves, and, as the day goes on, signal a "whoa" or offer with great good humor a gentle prodding, keeping us to the pace, in sync with the patient city workers who collect the bags we fill. Other neighbors contribute lemon water and, later, very generously, lunch. In spite of the loss of many traditional families in the Canyon Road area, values of neighborliness, pitching in, and wonder hold in the hearts and hands of these residents.

An energetic nine-year-old, Collin, gathers leaves for his father at their end of the acequia, lopes down between the rock walls, collects trash at my end of the section, hops back to his dad's bag of debris again, then back to mine. Focused on the job, he hasn't said much until we chat about a few worms we've found together. Piling even more leaves and branches into our bags, he stops, looks me squarely in the eye, and confides, "I don't know what I was expecting, but this is funner than I thought it would be!" It is fun, a savory lot of fun, a good workout, a community task. A mixed group—New Mexicans, Texans, New Englanders, Californians, a Coloradoan/Quebecker, people with querencia rooted here or elsewhere, some finding it for the first time—we weed, clean, rake along, stuff and pile bags, lift, chat, laugh, and sweat in the Acequia Madre. Narrow, like all the acequias, but Santa Fe's main artery of irrigation nevertheless, the "mother ditch" carries the water through the neighborhood, into its laterals, *acequia Analco* and *acequia ranchitos*, and on south through town, bringing "the beneficial use"—as the law states—of water to needy gardens, trees, and the few farms that remain within the city's perimeter. The state engineer's office sets the priority date for the Acequia Madre's water rights at "time immemorial and prior to 1680." Working between its banks in 2007, close to its ground and dust, I am even more aware of how vitally this ancient tradition revives kinship even amid our tensions and conflicts, and links us always to the mountain, its precious snows, and beyond to the larger, in some parts, still empty, uninterrupted land we know.

Between the rain and a sudden hail shower, there is a quiet, pithy jubilance among us. The ribbon of runoff tumbles steadily down from the mountain toward the two small gates. We watch. The time has come. Down, past the trees, toward us, the crystal water flows, soothing our cells. Reaching the shore where we stand, it branches under one already-opened gate, flowing into the Santa Fe River until the children's pulling and turning on the acequia's gate redirects its course. Was Collins's delight any greater than our adult joy as we watched the children being shown how to open the gates under Phil's steady hand and buoyant flash of smile?

Release comes, the water, trickling first, then splashing into the Acequia Madre, our acequia, spreading into a deep hole in the dry bed, gurgling, then free, moving now toward the culvert and through it, alive on its seven-mile course. Good water. Beautiful ditch. We watch, then play again as Phil opens the sluice gate wider onto the Santa Fe River for a moment to show the kids the rush of a falls. Querencia!

The same thrill of water had come to me years before in my studio along the Santa Fe River. Occasionally while I was sculpting, a tumultuous rush of water would rise and pound outside the window during a thunderstorm. Energized, exhilarated, enveloped by the ozone, I would stand and take in the sight of the sudden rapids cascading by in what was only moments ago a dry riverbed. The water was especially reassuring in those years, the 1990s, as it was too often shut off at the reservoir above, and cans, wine bottles, and other litter became as common as water might have been. It was a time when storing the runoff from the snow in the mountain reservoirs was considered good water management for an arid climate.

Sad vegetation, tattered and eroding banks, lack of water itself, bulldozers flattening the shores, the all-too-common trash are familiar to us in the Southwest, and are all signs of a habitat under duress. When we amble along beside a dying river, we are cut short, disheartened by the imbalance, the sometimes overwhelming physical sense of neglect. Naturally these realities, more than signs, undermine our courage and our connection. Richard Louv relates the naturalist Robert Finch's thoughts on this:

There is a point in our relationship with a place when in spite of ourselves, we realize we do not care so much anymore, when we begin to be convinced, against our very wills, that our neighborhood, our town, or the land as a whole is already lost. . . . the local landscape is no longer perceived as a living, breathing, beautiful counterpart to human existence, but something that has suffered irreversible brain death.

This is too often true. In so many places, we have brought about a place devoid of the moving holy. Our water, our very food, our spiritual sustenance depend on our restoring our affection for the land. As much as we live with the risk or the actual experience of disaffection from the land—and we do—we can still take heart in the places, traditions, and symbols where we can restore both our affection and the earth. As precious as rivers have been over the centuries to the land itself and its animals and peoples, a living, flowing Santa Fe River—source of irrigation, of drinking water, of beauty—is our palpable heart line.

On a bright Friday afternoon, as the February snow melted and softened the clay earth outside, Dominique Mazeaud welcomed me into the warmth of her straw bale adobe home where I had come to hear more of her story. "Passionate, it was a passion, like a love relationship," she repeats with delight as she recounts her childlike rapport with the river and her enthusiasm each time she rode her bike to its shores, dedicated to its cleansing, curious about what its offerings would bring that day. The river needed attention, the intention of her ritual.

Profoundly distressed at the sight of the trashed and sorry state of the Río Grande on a visit to Taos in 1986, Dominique had been moved to make a pilgrimage of sorts. For seven years (from 1987 to 1994), on the seventeenth of every month, she walked the mostly dry bed of the Santa Fe River, a tributary of the Río Grande, cleaning it and blessing it with ritual. An artist already committed to the spiritual in art at the time, she made this her artwork: a ritual performance, a witnessing, a healing.

She speaks of the first sadness of seeing the pollution, the neglect, and lack of life in the riverbed itself and amidst the natural debris, its banks trashed with dildos, discarded furniture, soiled blankets, of course the myriad cigarette butts, and the needles reflecting the drug use along its dead shores. But the river offered gifts as well. A Jesus statue also surfaced, and a copy of *Black Elk Speaks* "coincidentally" appeared out of the ruin on the day a Native American teacher had brought some students to join in the cleaning. "Magic, the river was magic," she repeats, enjoying still the significance of this synchronicity. One day a Hispanic man approached with, "You want trash? I got trash," and offered to show her serious trash in his yard. This encounter led to cleaning first, of course, then to the planting of a community garden for him and his friends, who happened to be part of the drug culture.

Magic too were the shoes, numerous shoes, starting with a little pair of Mary Janes. Such an appropriate artifact for her story, as she tells of the playfulness in the bending and bagging, the hopping and climbing among the banks, enlivening within her a whole new childlike spirit.

Reading excerpts of Dominique's journal, I am reminded of Aboriginal song lines when she says "partaking of the river's feast . . . I am simply dancing the song of my heart." Suffering and exuberance, defeat and healing. In the aboriginal song lines, as in the Inuit hunting songs, land and human echo each other, the songs mapping the trees and boulders, shelters, dunes, or ice floes guiding the people across known and un-known territory. Dominique's monthly ritual created a new heart geog-raphy for her, for others. Her act, her *song*, reminding us all: the water needs our care. Her ritual art wove the love of the river with those who occasionally joined her as she repeatedly did the necessary bending and bowing, an unusual kind of genuflection. Jay Griffiths, in her book *Wild: An Elemental Journey*, quotes the aboriginal painter and story-doctor, Margaret Kemarre Turner, who also knows the reciprocal power of place: "Going to a place keeps it alive," she says, "and keeps you alive too." This was happening here on Dominique's solitary walks. This we know—in our own heart, our own breath—as we too walk along. This happens as, our skin refreshed, our eyes rested, we meander with the current.

Dominique followed the dry riverbed before water engineers estab-lished that 40 percent of water is often lost to evaporation in the reser-voirs; before environmental science and practices in Australia and South Africa, as well as Texas, Colorado, and Arizona, renewed understanding of the importance in keeping the water moving for humans, the ground-water, aquifers, fish, and the riparian habitat itself.

This artist walked years before the current Mayor Cross committed to restoring the river—still one of the most endangered in the country—before letting it run at regular intervals as a living entity; before he and Youth Works in 2001 developed the Santa Fe Youth Corps, engaging youth from all backgrounds to help rid sections of the river of trash, up-root invasive species, design conservation measures, and do some of the plain but intelligently designed hard, necessary work with boulders and rocks that naturally reinforce the riverbed, slow down erosion from storm runoff, and create elegant, healthy, new meanders.

Dominique's pilgrimage ended in 1994, as more water entered the river and the work became lonely, even a little frightening, the farther south of town she went. Thirteen years later, on Earth Day, the city convened the community at large to participate in planting fresh cottonwoods and willows along its banks. So many showed up. Men, women, and children energetically rolled up their sleeves, heartened by intergenerational efforts, soothed by the aliveness of the restored sections of the river. Days later, walking alone along the newly graceful shores of the river south of town, I, like so many others, hoped for the continued flow of water.

Again this early spring, another friend relates the nimble joy of children scampering up and down the banks picking up the trash, and her own pride at how pristine her section of the river becomes. In our dry climate, it is a keen pleasure knowing that the water about to come will seep in, not just under the riverbed, but will nourish the land and groundwater up to hundreds of feet on either side. "Nurture the country because the country will always nurture you," Margaret Kamarre Turner reminds us again.

Wholly engaged in healing the river—the humble, dirty, delightful process—Dominique's art ritual began with what she called the "matter" of "trickle trash, disrespect, and despair." With repeated gestures of her ritual "work as a garbologist," she says smiling, listening to the beauty of the river even without water, she brought it alive, healing herself as well, and along with her the people she encountered, all the while moving our culture toward repair and a finer knowledge of the land. I can't help but see this pilgrimage as contributing to a kind of continuity in the river's life and the culture around it: a buildup toward a continuum with Dominique's solitary commitment to river care eventually engendering that of the city's—and the larger community's—care and concern.

The awareness of place gained in paying attention, touching and cleaning this narrow bed of earth goes beyond its immediate banks. "Standing in the riverbed, the vein of the earth!" Dominique exclaims, evoking for me the weaving through the land not only of the Santa Fe River but the Río Grande too, and the Colorado, on through their tributaries and up to their headwaters. In the bright intensity of her dark eyes shines the visceral knowledge, an ancient one, as she reflects on "our holy river . . . the whole American network of rivers. Soon a diamond web stretching over the Earth."

In the care of the acequias and the rivers—these "veins of nature and wilderness," as Richard Louv too puts it—we carry on our species' deep-seated knowledge of a physical, communal, and symbolic place where we can love the "circumstances of our existence," reviving our affection for the land, putting it first over our habits as consumers and believers in pursuit and progress. Losing beautiful wilds of our continent, we are forced to realize and actually believe that the earth's continuance and our own are intricately knit. Refurbishing our daily experience of beauty, nourishing our communal bonds, the vitality of moving water sustains our future with the land that remains.

TWELVE

Paradox

Participating in the 397th cleaning of the acequia carries a personal irony for me. Several of the new people I meet on this morning are people who arrived to buy their million-dollar Santa Fe home in 1993, the year I left. They are the newcomers whose wealth precluded my being able to stay home. Fourteen years later, here we are joined together in the pleasure and maintenance of the acequia in its vital role of restoring water to the area, and querencia to the community.

Paradoxically, the painful economic realities I had known still hold. Extreme wealth in western real estate still has major impact on the rural and low-income people and the land, but in a new-century West our divides of wealth are not always what count. This is reassuring, a recognizable spirit of the generosity of the old West in the New. In a fragmented society such as our twenty-first-century America, our community relationships are often temporary, but the bond is alive and long-term, grounded by the few who can hand it down. Here, as in the Old West, a newcomer who is worthy of trust and capable of attention and respect toward the culture recently entered can become a respected neighbor. If they are still welcomed, it is thanks to the traditionally hospitable people like Eleanor Bové, who carry the steady kindness, the people savvy, the genuine respect for good people, and the quick wit and humor so warmly recognizable in the Indo-Hispanic culture when dealing with ignorance, spite, disrespect, or outright racism. Eleanor's own inexhaustible stories and appreciation of innumerable interesting people—artists, writers, musicians, politicians—who have come to Santa Fe throughout the decades continue to enrich respect and friendship within our human tapestry.

The acequia culture that she, Phil, and other parciantes and mayordomos watch over has meant survival and continuity to the farms and ranches, sheep and horses, towns and cities, and their people. It is alive, living, punctuated by the celebratory spirit of spring runoff despite the increased taming of the wild and the real estate assaults on the land and population. Hispanic, Native, or Anglo, the human heart recognizes querencia, *terroir*, land, *tierra bendita*. Our words from ancient agricultural practices connect us to heart traditions of knowledge and the necessity and naturalness of collaboration. They are not about the accumulation of land but about our affection and longing for it, resonating with the sense of safety in a place we call home.

The morning I walked up Acequia Madre Street to join others for the cleaning, there were dozens of cars parked along the narrow street, a highly unusual occurrence in the neighborhood. Definitely something besides the work in the ditch was going on. Elegantly casual people were ambling curiously down a side road to what I later learned was a major estate sale being given at the home of one of the more famous members of the wealthy neighborhood. The contrast was striking, like two cottonwood branches glimpsed in the clarity of desert light: the high-end consumption of Santa Fe style and the simpler, earth-related activity were juxtaposed along the ditch on the day of the 397th acequia cleaning. The ancient continuity of survival structure and our contemporary exaggeration of wealth, independent of survival needs, seemed to create a visible "tipping point." This clear juxtaposition was "so very Santa Fe," "so very New Mexico," perhaps, now that the phenomenon of wealthy arrivals has spread to Albuquerque and Taos and beyond in exurban development, even "so very Montana," "so very Colorado."

It left me not only curious but concerned about the fragility of the ancient tradition in today's urban setting. Phil Bové points out, "As some praise [the acequia madre], others will exploit her. We must keep steadfastly to our course of protecting the acequia. . . . Not a week goes by that we are not meeting with city staff or developers working out details concerning the acequia in future projects or trying to correct damage done by the unscrupulous."

Here in this neighborhood, the majority on both sides of the juxtaposition are among the highly privileged class economically and, now, mostly Anglo. Divides between the rich and poor remain; cultural tensions remain. The trouble in our culture now is not necessarily the desire itself for land but the ignorance of our own cultural fabric that could sustain us, coupled with the unlimited, egotistical rule of money that reveals the lack of respect and love for this land and its limits.

It is reassuring and, to some newcomers, astounding how strong a quiet culture can be. "The power of a gift economy," like that of the acequias, "remains difficult for the empiricists of our market culture to understand," Jonathan Lethem clarifies. "What's remarkable about gift economies is that they can flourish in the most unlikely places. . . . A gift economy may be superior when it comes to maintaining a group's commitment to certain extra-market values." Friends in the urban east ask me: If this sense of gift is viable, can the experience of it possibly be true today in a world so driven by greed and conflict? Yes, it can. It is still the essential element of the fabric of westerners' culture, where survival tensions and joy coexist at the heart of the community's rapport. Surface water, groundwater, well water, the aquifer. Numerous and sharp—like spines on a cholla cactus (ouch!)—contentions over water and grazing rights are fundamental to the *modus vivendi* that has evolved in the West, especially in regards to the grazing on national forest land or to oil and gas development. In spite of these difficulties there is ongoing pride among the people, because care and belonging between the distinct cultures have always existed. If tensions over water influence politics, culture, family, and neighborhood, the earth-related values remain as a core that organizes and influences people. This keeps Taos and Santa Fe as "cities different" but vulnerable too, due not only to limited water but also to the current thrust for natural gas and oil development.

In the Southwest, as in other complex societies, people enjoy one another and one another's culture. Montreal, for example, was and is first Mohawk land, then a city founded by the French, then by the English, then host, and now home to diverse immigrant communities as well. In *Translating Montreal: Episodes in the Life of a Divided City*, translator Sherry Simon writes of her complex metropolis: "The differences within a city give rise to interactions that fall along a continuum of mistrust, resistance, and vivifying exchange." The delicate balance of trust and mistrust, of tolerance and intolerance in the West has still allowed for a coherent identity because of the recognition of the essential bond with land and water. Caring for the nurturing beauty that the people live in brings a peace among neighbors, at once strong and careful, definitely rooted. The idiosyncrasies created by the shifting cultural balance have added to the character and independence of spirit in the people and insured that the reliable medicines of laughter and good humor are part of the vigilant guardianship protecting this Canyon Road neighborhood—like many others in America—from litter, noise pollution, and ignorance.

Others will claim that shared water, like actual communal lands, is a thing of the past. Fortunately this is not always true. In a decision upholding the Acequia Madre's water rights over claims for it by PNM, the public utilities company, District Judge Arthur Encinias stated, "The reports of the death of the acequias in Santa Fe are greatly exaggerated. The people, the land, and the water are intricably bound together and will be until Santa Fe is entirely paved over. It is this culture which is our greatest pride and not without considerable value, though not measurable directly in dollars."

Along the four acequias in Santa Fe, it is heartening to know, as the *comisionado* writes, "The remaining users are dedicated people who know and appreciate their rights. The acequias are aware of the rising value of water rights but have so far been able to resist the temptation of selling the rights for the high dollar amounts offered." We are learning again the dollar is not almighty. The physical place we live in matters. Without it, humans, too—all originally related through our DNA in spite of our incredible diversity—may be at risk of vanishing. If we pay attention, we cannot help but understand that we live on a rapidly changing planet whose imbalances are still powerful enough to collapse the biological systems that support not only the animals' but our lives. At a time like the present, when even the backyard honeybees are disappearing— their lack of pollination putting fruit crops as well as wild vegetation at risk—the spirit of querencia is essential. Although as I revise these lines in my Quebec garden, one bee makes her presence known, the absence of billions of her kind this spring can't help but bring home the awareness that the deep and intricate natural balance is shifting, that our food and the sweetness—the *deliciousness*—of our lives is at risk. Our divides can no longer be what count; the spirit of "gift" needs to be nourished. The solidarity that people in challenging lands have depended on for multiple centuries now needs to be activated, restored, and applied not only to places we used to consider as extreme—the arid desert and the cold north—but to the survival fragility of the earth balance as a whole. As some are doing in Santa Fe's acequia madre neighborhood , crossing our own divides of rich and poor can allow fruitful work for the earth.

North America is not replete with ancient continuities. Las acequias are in great contrast with our general culture of drive-throughs and drive-bys, where people rarely carry a daily understanding of where they are or a shared, tacit sense of joy and pride at being who they are in that place that shapes them.

Holding on to or recreating community symbol and celebration of our need and affection for the land can help us if we are to resist the polarization of a two-tiered society and resist the confusion and conflagrations of a global world that disperses identity. Our heritage of the Western Spirit, of our querencia—a way of life enjoying reciprocity with nature—carries the resilience we need. Passing this heart knowledge around and passing it on to the children offers hope for establishing a new inheritance for our country.

I know how deeply true this is as Eleanor, a lilt in her voice, tells me, "I want them to love it." She is kind to children when she finds them playing in the acequia, simple teaching and kind teasing being more effective in helping the children respect the "bed for the water." She knows; she grew up on the acequia, too. Generous with their knowledge, Eleanor and Phil, Estevan Arellano, and other committed people like them do all they can to ensure the passing on of the tradition.

Due to development and recent drought, pressures on the state's water supply have led to helpful innovations. In Taos, through the *Sembrando Semillas* (sowing seeds) project coordinated by Miguel Santistevan, young people have been saved from cultural confusion and drugs by bonding with their elders and their heritage through the acequias. Farming their own gardens, improving their food security, survival, and joy, they learn firsthand—and also witness in the videos they are encouraged to produce—how much the land and its people require each other.

O New Mexico

You are being gentle again
with the sudden sight of water
> *trailing*
> *high in an acequia above the road*
> *through softly mounded*
> *banks,*
> *tenderly raked*
> *rounded, embracing water*
> *barely springing new clover*
> *narrowing toward the sun*

O New Mexico
you are being gentle again

rippling at roots,
moisture flowering at the foot of trees   leafing
lapping under grids and small bridges
the acequia—dry earth and grass consenting
    to water   rippling

Affection for prairie or mountain does not lead the way in North America, where whole populations of youth say they connect more to their computers than to nature. That essential connection does not necessarily come automatically, even when some of the children live among vast beauty with forests, farms, and sage plain to roam in. In some communities, the respectful and celebratory spirit toward the land is alive, but in others, if the beauty of the wild and the responsibility toward it is to hold, as my dad knew, "You've got to be taught." And encouraged, for there is much to cherish.

Backyards still hold discovery. Entering one from the open land bordering the Río Grande, Sarah, Charles, and I find their sons continuing work on their backyard construction. Familiar with the land from family hikes and their first years living in national parks, Caleb and Sam are clearly at ease as they wander back onto the trails, soon to reappear with recycled wood and materials they have gathered, all of it exactly appropriate for their purposes. Building their two-level fort—their version of a tree house in a land where there are not many trees—I watch them and their friends ponder the angles and edges, finding the right fit for a new board. Confident. Independent. Their senses, imagination, and skills are clearly enlivened and honed. Trusting, too, they consult with Charles who keeps a discreet distance, intervening only when called upon. Once the building solutions have been found, the adults prepare dinner, watching the boys enjoy a game of basketball with their buddies. In the comfort of friendship the evening meal ends. The trails beyond the yard lay quiet as they meander toward the Río Grande. The backyard is enlivened with camaraderie. We cannot hear the river but, embraced by the land, we know it is there.

THIRTEEN

Homeland Dilemma

The hawk soared
the eagle cried
songs fluttered
the sun rose

We are necessary to the sun
and the rain?
that too

In his collection of autobiographical essays, *Owning It All*, William Kittredge summed up the mythology of western Americans in these words: "In short, we see ourselves as a society of mostly decent people who live with some connection to the holy wilderness, threatened by those who lust for power and property. We look for Shane to come riding out of the Tetons, and instead we see Exxon and the Sierra Club. One looks virtually as alien as the other." In spite of its desire to protect land, the conservation movement has not often understood westerners' connection to the sacredness of the wilderness and ranchland.

Perceiving ranchers as self-interested and abusive to their wide acres, conservationists have often held little respect for the knowledge, love of the land, and traditions that ranch culture carries. The bone of contention is that, in the view of ranchers, farmers, Indians, and other old-time residents of the West, people are important to the land, whereas a majority of conservationists have privileged the view of the wilderness as sacred only if devoid of people. This divide still exists. Fortunately, many are trying to bridge it. In *Working Wilderness: The Malpai Borderlands Group and the Future of the Western Range*, Nathan Freeman

Sayre brings us voices to heal the divide and deepen a necessary understanding. Whether our affection for wilderness is based on our sense of it as an unspoiled, primal world or as a place of family living, learning, and growth, "in both cases," Jim Corbett, the innovative rancher and cofounder of the Group, states, "'love' means valuing the land in itself, and this is the foundation for establishing basic rights for native biotic communities." We humans, then, are necessary to the balance. This ethic of reciprocity is at the heart of newer conservationist efforts to repair the land.

The intricate knowledge of the beauty in the still-existing wild is what ranchers and farmers, native communities, national park or national forest personnel, and residents know to be essential and precious to their lives. The beauty is one and the same with the life balance among these people who know that scattered, isolated pockets of wilderness cannot adequately ensure continuance for the finely woven biological processes of life there. These are the same people who have known that the loss of the empty space in our wide land is critical to us and the animals before studies proved that "tree museums" and smaller parks are not diverse enough to sustain migration and habitat. These are the people who know much sorrow or bitterness at this loss. Whether on a working ranch, an Indian reservation, or a temporary homesteading situation, what Jim Corbett calls "errantry" is the movement between people and animals "that is primarily concerned with communion, which in our age focuses on the harmonious adaptation of human civilization to life on earth. The first decisive step into errantry," he recalls, "is to become untamed, at home in wildlands [where] one must accept and share life as a gift that is unearned and unowned." A gift. Our roaming the West taught us this.

Just beyond the bridge on the Colorado River, I open the gate for my Quebec friends who have come to share my last week of homesteading in the Kawuneeche Valley. It is 1995. I have a few more days left to myself at the cabin, but at the moment I stand immobile, welled up with sorrow at the approaching finality that their departure signals. Having experienced the valley as the *premier matin du monde*, "pristine as at the dawn of the world," they too are grieved, knowing what the closing of the gate will mean. Intending to console, Jutta speaks. "You carry it with you, you carry it with you," she repeats, her voice soft and kind. Yes, we carry it with us, we carry it within. But away from the land we are powerless to care for it, powerless to love it in a practical manner. This is a kind of exile. We are not there, *in situ*, to accompany the rushing of the river, the elk, and the brilliance of the hummingbirds in the morning. No one will live so closely with them; they will be powerless to restore us. Here and elsewhere the land is widowed, and our continuity of at-homeness, of being "untamed" is gone. As I say this, I wonder if I too suffer from that hopelessness for the land that Robert Finch spoke of. We are all, after all, part of the fabric of our culture.

The rhythms of the Buffalo Dance, a finely honed silver and turquoise bear claw necklace, the smooth surfaces of smoke-fired pottery, and the vibrancy of more contemporary paintings and sculptures are often vehicles of zoë-life, carrying the recognition of the spiritual in a community. The family cabin, the Pueblo plaza, traditional hunting grounds—the actual physical places that give us our stories—are even more deeply resonant and very often the origin of this spirit life. In the vast landscape of stone and ice in the North, the Inuit speak of *uganatug nuna*, meaning "a deep and overwhelming attachment for the land," Norman Hallendy (a close friend of theirs) tells us. For just over a hundred years, national parks and monuments, their geysers, peaks, rivers, meadows, cactus, and forests, all wild and rich, have afforded access to such a source of peace for countless Americans.

In 2006, a celebratory reunion took place at Mesa Verde National Park, one event marking the one-hundredth anniversary of "the Mesa." All those who had lived in the park since its beginnings in 1906 were invited to gather for a weekend. And gather we did, arriving from all corners of the country, from Australia too, and Quebec. The silent cliff dwellings, hand-and-toe trails, red rock walls striped by desert varnish, the juniper forests, and ancient pottery were the common ground that drew us. It was a simple joy being back among fellows and the presence of the "old people" who had built the dwellings and farmed the mesa tops hundreds of decades before us.

Archaeologists, rangers, maintenance staff, volunteers, administrative personnel, we made our way the first October night through the mountain cold under the crisp black canopy of stars to the warmth of a meeting room where a storytelling evening was scheduled. Several men and women on the stage, who had actually grown up here, regaled us with tales of childhood pranks, favorite horses, the building by the Civilian Conservation Corps (CCC) of the now-cherished visitors center, the first southwest archaeologists' restoration efforts—and mistakes!— in the ruins, and throughout, the kindness and tenacity during fires and drought, the freedom and care insured by an extended park "family." Their stories mingled with ours who were still working (or had been) for one or twelve seasons or a whole career. Laughter and silence rang throughout the standing-room-only crowd, echoing the genuine recognition and renewal of our belonging to this West. More than nostalgia, more than anecdotes, this was one seamless story of belonging to a place, "the Mesa" being home to all of us in one degree or another. Our shared delight evoked better times we all had known, when the parks were a force, equipped with adequate personnel and Theodore Roosevelt's belief that protecting "the majesty and beauty of the wilderness" for the people "was a great moral issue and integral to the continuance of the nation."

Another evening during the reunion, old friends and new gathered at one of the beautiful old stone-and-viga CCC homes for a potluck supper. Amid the jovial conversations, I asked one of the few career rangers left what volunteer job I might do if I came back for a summer. In response, she grimaced then retorted, "Superintendent!" After the initial surprise, smiles broke into laughter, shoulders relaxed, everyone recognizing a good joke on the widespread, current despair in the parks, welcoming humor, a good antidote to the loss we all knew. The truth is, in keeping with the federal policy not to replace permanent positions when they become vacant, the loss of qualified personnel in the parks is acute. The few remaining stewards of these lands are in distress. My friend's exasperation imagining a volunteer as superintendent stems from the sore reality that now many parks are being run with a majority of untrained volunteer personnel. Since the beginning of the new century, fire, drought, beetle infestation, floods, and increased crime in some parks have complicated the task for far too few hands.

Volunteers have been an integral and highly appreciated part of the Park Service since its founding, provided by its own Volunteer in the Park program as well as a private program, the Student Conservation Association. Whether during the Depression or the prosperous years of the sixties, traditionally the rangers would educate this helpful labor force in the spirit of the park or monument's mission, while the volunteers found the satisfaction of sharing temporarily in the proud stewardship of protected public lands.

I was fortunate in the 1980s to be a volunteer before major budget cuts and shifting policies in Washington affected the National Parks' infrastructure. Documenting petroglyphs and plaster conditions in the ruins for Mesa Verde's archaeology lab, my purpose, like that of other volunteers at the time, was to supplement the rangers' work. Not only did I have lodging in the simple homes Ernie and I had shared, but the Park offered me a studio space—an empty office—where I had time to savor *uganatug nuna*, as I contemplated the radiant but stark elegance of the desert and mesas spread before me. Amidst the warmth of the one hundredth anniversary celebration, I was thrilled and honored by the pleasure former colleagues found in the paintings I had made in that studio, which still hung in their homes. Our reunion conversations revealed too, in situ, the extent to which government policies had undermined the quality of the volunteer tradition. With the major reduction in qualified personnel, rangers must often supplement the work of untrained volunteers, which does not make the job easier!

The spring following the Mesa Verde celebration, over a decade after the Grand Lake gate had closed, I join several park-related friends arriving from a variety of regions for a hike. Delighted to see one another, we fall into step and wind our way among the glistening ponderosa pine of the Santa Fe National Forest, up toward the altitude where aspen groves cover the slope. The day is replete with familiar smiles, hugs, and laughter, joy at the spring undergrowth, as well as faces marked by worried brows, voices resonant with disappointment, and the predictions of more loss.

The strain of actually participating—through being responsible for managing budget cuts—in the demise of the national parks they have lived and worked in for years wears visibly on weathered faces. Amid enthusiastic or concerned news of children, keen interest in one another's health, or ramblings about vacation treks to the Grand Canyon, a litany of laments surfaces and intertwines. The consequences of the lack of qualified personnel, independent of the park in question, weighs heaviest in their hearts.

"It's not a community anymore!" Bob exclaims, shaking his head. "Volunteers and rangers don't have lodging inside many of the parks anymore. There's no heart and soul left, no core of people who know the place." Glad that I was a volunteer in the days when we were housed, given a small food stipend, and welcomed as very much a part of a kind and stimulating community, I listen to stories of well-appreciated, motivated volunteers being solely responsible for backcountry patrols, filling in the gaps for interpretive tours, archival work, scientific study, or other necessary positions, carrying responsibilities they are not trained for.

"In spite of their goodwill, though, the public's safety is at risk, the resources, too," Sharon elaborates, "'cause volunteers are not required to meet the criteria for skills and knowledge demanded of rangers."

Dark-haired Marie interjects a note of hope. "Micromanaging from the bureaucracy is terribly frustrating, I agree, but in our park we still have an adequate number of rangers who've actually known the valleys and forests for years. Among the volunteers, we've got two different couples, retired people, who've been coming back for a long time since they're able to provide lodging with their own RVs. They're friends now, almost like permanent rangers, knowing the rivers and trails and offering really competent administrative and backcountry service." As Charles frowns, she concedes, "Some of us are luckier than others."

"At our place," Kurt intervenes, "there are only three backcountry rangers left—down from seven!—sometimes only two if one is away on training. We almost never get to the backcountry to repair trails or bring supplies. Nope! Can't afford a day out there with all there is to do in the front country. On our days off, the fire crew and rangers remove fallen trees and do other more specialized maintenance. That's not our job, but it has to be done. Even with that, we're way behind."

A pensive group, we walk along, the sun on our faces, no one having another joke immediately available to break the silent acknowledgment of the impact of the changes. Sue, a nurse who has worked with Pueblo and Navajo communities, points out the irony that "even though born of tragedy, some reservations have helped preserve Native Americans' access to their birthright."

Tor, Sue's husband, and I are especially saddened by the loss of the horses. In some parks, like Yellowstone and Bandelier, stock use was historic. Carrying heavy signposts and diggers as well as skilled rangers on their patrols, the horses were necessary for trail maintenance and supplying research and other field activities in the varied and difficult terrains. Without them—and their packers—many trails can't help but be abandoned or downgraded. Beyond their practical use for the parks and the visitors, the horses were also emblematic of an ongoing relationship with the wild. We mourn the animals, absent now from these landscapes, receding from the public's imagination, removed from the care of the parks' staff.

I listen to my friends as tiny blue and lavender flowers wave in the alpine grasses, wondering if the new corporate models will eventually be able to relate to people and the land. The gap is wide right now. What the old-time rangers deplore the most is the imposed neglect of the trails, many of them historic, built during the New Deal by the CCC. Deer trails, ancestral hand-and-toe trails, Ute hunting trails, national park trails are all threads weaving us to the wild. Paths of wonder and knowledge, they are the essential links to what the wildland has carried for us, given us, taught us.

Kate, a career ranger's wife, raises a commonly held concern as we take a break, seated on the comfort a few boulders provide, a deep valley below us. She recounts the discoveries made on a recent hike with a friend who introduced her to an old backcountry trail. Describing the lush canyon they meandered through, the abundant waterfall they reached, the fascinating traces of humans gone before, she speaks earnestly. "This river trail is so beautiful! Why didn't I know about it? What are we missing now? Who can tell us, show us, teach us about these treasures! Why don't the people at the visitors' center or the younger rangers know about this trail anymore?" Aware of the safety risk to the public, her dismay had been piqued by the disheveled condition of the trail and especially by the worn signage, askew or fallen down. Reminding me of Walt Whitman and his wondering who would speak for the earth, "bind us to it," she answers her own question: "It's as if it's not so important anymore. Why?"

With one voice, the rangers react: "There's no more money for seasonal trail crews in the park!" "Without skilled ranger staff, emergency trail repair is delayed—and piecemeal when done!" Another complains: "Routine trail repair isn't happening at all."

Keenly aware of how budget rationalization has created a void in place of refuge, learning, adventure, and community with the land, we fall silent again. The breeze is gentle about us as we trundle down the slope toward our cars. Bob brings us back with a "Duty and my teenager call!"

Lingering goodbyes at the trailhead provide time to savor our shared reverence for the canyons and peaks we have so long worked or wandered among together. I am moved to gratitude for these friends, committed stewards and their families, dedicated to educating and protecting our, the public's, rapport with the natural world and its inhabitants. They have facilitated our access to a common home. Living the loss of their work and Theodore Roosevelt's vision, facing an institutional disregard for the public and the original mission of maintaining public lands *for the people*, they have held on against the odds.

On the way home, Tor says he is concerned about a new focus in some parks: their website. Perhaps it is a good step beyond the more familiar drive-by visits and a necessary tool in our age but the web is hardly a place for direct experience. Money and effort are being put into websites with some administrations and policies, he feels, privileging the virtual experience for visitors over direct contact with the forest, the river, the desert. Removed from the actual experience while on the Net, neither children nor parents nor solitary explorers can find places to breathe in the freshness of pine; enjoy the invigorating mountain chill on the skin, relax after the sweaty, satisfying climb up to timberline; or sit silently rapt on the very stones placed there by fellow humans a thousand years before. It is our bodies that know the beauty of the land and the sea.

Needing an optimistic example, I remind Tor and myself of the thirty-some-year-old program at Florida's Biscayne National Park. There, city kids more familiar with drug dealers than cooperation learn to fall into one another's arms to build trust and work as a team while learning from the rangers how to paddle canoes through the bay themselves. Skimming across the wet, sparkling habitat, they marvel at sea grasses—meadows under the water!—blue crabs, seahorses, and the flamboyant fish living "down there" among the coral reef.

The national parks and monuments are where most Americans who were not on reservations, ranches, or farms could experience the sacred, enlivening rapport with the wild. If they, like my relatives, trusted the parks and were able to enjoy, learn from, and know the intimacy with the wild for a century, it is due to the vision that Theodore Roosevelt instilled in the hearts and minds of park personnel and the public. As in-holders, my family and our neighbors collaborated with that view in the spirit of custodianship and caretaking, as people living simply on the land, enjoying and nurturing its chemistry of beauty.

As the park culture eroded from lack of funds and suffered—as from a vitamin deficiency—a "personnel deficiency" that higher fees and more complex regulations cannot make up for, we, the public, have risked being left with only limited vestiges of land held for the common good, a sorry state of affairs that reveals our government's propensity to orient its purposes elsewhere, often to wars and oil exploration and exploitation, against the hearts and minds of those who know the wild and its importance.

Witnessing the sense of painful disinheritance, I have wondered if it wasn't time, once and for all, to grieve for this institution born of Teddy Roosevelt's love of America's wilderness. So many serious questions remain: Can the project for a uranium mine two miles from the Grand Canyon Visitor Center be stayed? What will the fate eventually be of the pristine expanse of the Arctic National Wildlife Refuge, its caribou and people still being defended from oil development? Can the exploration and drilling leases in or near there and other park boundaries definitively be revoked? What adaptations will be made under the pressures of energy development for new populations? Finding a balanced relationship in managing and protecting the parks from the industrialization and urbanization of the West is no small order.

Avoiding an equally painful answer to the thought of giving up, I tap into the grapevine, finding good news instead. During these years of duress, other resourceful, passionately committed individuals from the national park's culture have insisted on digging in their heels and believing in the parks' true mission. With limited or no park budgets, they have sought support and developed projects so that these sacred places stay alive in our imaginations and our affections.

A good friend and career park employee has steadily insisted that all is not lost, citing the Vanishing Treasures program initially started by a handful of park managers who were aware that insufficient funds were as much a threat as erosion is to the Southwest's archaeological resources. In place since 1993, this collaborative effort is committed to renewing the preservation workforce, saving cliff dwellings, churches, and forts from deterioration.

The National Park Service Retiree's Alliance has been fighting hard for maintaining and renewing adequate numbers of personnel as the only way to insure that resources are properly managed and visiting children and adults given safe access to the natural and historical riches the parks hold.

In *The Mesa Verde World: Explorations in Ancestral Pueblo Archaeology*, the School for Advanced Research in Santa Fe recently published significant research by a number of dedicated southwestern archaeologists, giving us a clearer understanding of how the Pueblo ancestors of this area lived. Soon, though, I, we, dare believe these efforts and others like them will be less isolated. And politically since 2008, with the Washington administration's renewed commitment to the parks, there is profound change and a sudden new confidence that the initial mission of protecting the land unimpaired for future generations of Americans can survive. Hope has reentered the larger fabric of national park culture.

For now, the Southwest has breathed a collective sigh of relief as—pending more environmental review—oil and gas leases on 132,000 treasured acres of golden rock beauty near Utah's Arches and Canyonlands National Parks, Dinosaur National Monument, and Nine Mile Canyon have been indefinitely suspended. In Tor and Kurt's parks, there's huge satisfaction at having a dispatcher hired again to maintain communications both with the backcountry and to the public.

In hopes of bringing youth back into the wild, several parks have been creating innovative new programs attractive to today's computer-oriented children. Dayton Duncan's and Ken Burns's PBS film, *The National Parks: America's Best Idea*, shows focused and enthusiastic Las Vegas schoolkids observing various colors in the sand samples they have taken with their own hands in the Ubehebe Crater. This is geology, this is hydrology. They are learning that land carries not only our memories but its very own, most minute and most grand. For this viewer, their pride is poignant, palpable, fun loving. In the arid, star-filled vastness of the Mojave Desert, a premier camping site, I am relieved that these children even have time, a rare chance, to be at ease with the sound of silence.

There is room to believe again in our shared values of beauty and health in the wild, and our link with the petroglyphs that trace our human past among the stone. Room for restoration too, because new funds allocated for infrastructure projects mean repair and can include, at long last, the opportunity to clear the debris, fallen trees, and mudded stone from the neglected trails that have weighed on the rangers' minds and blocked public access in the parks and on national forest land. Room for hope that the "culture of fear" and "ethical failure" the Department of Interior's inspector general began acknowledging back in 2004, be healed. And more than anything, hope is revived that more adequate numbers of qualified personnel will again be repatriated to the purpose of education and protecting the air, water, and forest for the people's safe use. Hope that, with them, we, the people, can reclaim our experience of wonder and "communion" Corbett speaks of among the woods, waters, peaks, and views that restore us.

FOURTEEN

Wounding or Pruning?

I was very moved, my darling, as I wandered
bright pink, with the yellow butterfly, through the purple aster
—stars, mon amour—down to the stream where there are more rocks
and the sun sheds a cloud

The calm in the breeze is unthinkable, mon ami, a kind of expanse
among the self-pruned trunks of Aspen, supple and velvet.
These trees are the eyes and grace of my mountains.
The full distance of space is here
Everything has breath, more sight, the same light.

The spirit of querencia, its refuge for the sacred, its kinship, its *inclination* to home, runs deep in our land, in our species. More than ever, we need to count on it for our regeneration as we face the imperatives survival imposes on us. In this twenty-first-century West, we acknowledge that paving, overbuilding, overgrazing, or otherwise exhausting the land has led to loss of food sources and self-determination for the community— indeed, of whole regions—and increased illness for people and animals, not to mention the loss of villages, water, and wonder. Our country cannot afford to lose any more of the breath and beauty of our empty spaces. The dangers remain for the West as it lives on in name and image, and fortunately in the real world, too. We have inherited a threatened land, but it can be reimagined.

William Kittredge saw the West of 1987 as being "snared in the un-certainties of a transitional time." The word *uncertainties* jumps out at me as I read. It seems so mild in this time when even the immediate future is unpredictable in the face of the worst financial crisis imaginable. Suddenly, we have been shaken by the overwhelming effects of rampant greed. West, east, north, south—we watch our own, more personal worlds totter along with the global economy.

For years now, the evidence of unbridled development in the West has been more than frightening. Real estate, the newest major alien power, came rolling "out of the Tetons," as Kittredge had said, and out of the Rockies, like a dark and forbidding storm. The justifiable fear of its power has been tangible on a daily basis. Actual and impending development has continually been funded and politically approved, independently of the availability of water resources.

"Snared," yes, we have been by the greed, our blindness to it, and the will of those who drive it. In spite of the housing crisis in the United States, it is doubtful whether the impetus to overdevelop will disappear soon from our cultural fabric. In the interim, though, we can begin to shift our values, release ourselves from the snare. The voice of Jim Corbett reminds us that creating a working relationship with the wilderness through our adaptation to it rather than our impositions on it is "a practical and spiritual calling." We are called.

Westerners, especially ranching and farming families I have met over the last few years, lament the ignorance about the land behind its fragmentation, criticizing our now patchwork landscape as an unholy rapport redefining land use in our country. Although some landowner families are accused of selling out to the irresistible millions available in real estate or oil and gas drilling, others do resist, staying on their farm or ranch, restoring its land, and often protecting it from development in perpetuity with what are called conservation easements.

At a city council meeting in Arizona recently, I watched a young developer respond to an elderly member of his community who had pleaded for protecting what remains of the desert in their area. Shaking his head, the young subdivider repeated casually, "It's just not gonna happen"—shielding, protecting the desert—"it's just not gonna happen." There it was, incarnate on my computer screen, the belief that Robert Finch identified, that nothing can be done; a belief that, even unwittingly, implicitly laces our attitudes toward the earth under our feet.

In Arizona, New Mexico, Colorado, and beyond; in Toronto, Quebec City, Barcelona, and Rome; and as far as the elephants' habitat in Kenya, the subdivision maps lay out new miles to be built up. In the Southwest, hundreds of developer licenses are approved monthly. Water is transferred, legally or not, from private wells on ranchlands, farms, and semi-rural neighborhoods to feed urban and exurban growth and the coffers of individuals who control the phenomenon. Ironically, while thousands of foreclosed and uninhabited homes await their fate, slowing the thrust of the economy, the land they sit on can't, won't, be reclaimed for food production, or for the new children's experiences of natural habitats.

How sad, this divorce—or is it widowhood—from the land. Familiar, yes, but still poignant. The few remaining places that are shared and protected collectively are so precious, the customs so rich, the disappearing elders so necessary to the passing on of the tacit knowledge and traditions of the community. As productive members of rural or ranching communities move to cities or at least seek part-time jobs to survive on the land, we admit how fast and complex our new century is. We have been used to diverse populations—sheepherders and other workers on ranches have been coming from Peru or Europe's Basque country for generations—but the University of Colorado's Center of the American West reminds us the "most telling reality" of today's West "is its new social layers": the prospering recreation and tourism industry, a lively cultural industry, and high-tech economy bringing new jobs, migrants, retirees, and others arriving faster than anywhere else in the country. Urban culture is embedded; as trout streams tumble by, espresso bars dot even the smaller towns. Newcomers abound in the cafés, near golf courses, and so do we, the old-timers. Even in the "old days," my parents and I would enjoy driving from Denver to Evergreen, a once-rugged but welcoming mountain town, for a meal or tea. However, as in many communities with precious rural resources, there are few locals who are not ambivalent now, or angry, about the sheer number of people encroaching so constantly on the foothills.

Can our consumer society learn to self-prune, to hone down our habits all the way to simplicity, suppleness, and strength so that we might protect the new ground for new life and a more cohesive culture, a more stable future? Alone or with those close to us, walking the trails among the green and golden trees in the Rockies gives solace and insight for our pain or powerlessness in our homelands' dilemmas. Aspen trees do self-prune, which is what gives them those "eyes" that gracefully sway in the wind and follow our path as we hike or ski. That's how they grow tall, straight, and strong to reach the open air and sky so that their canopy of quaking leaves will protect the necessary conditions of shade, light, and moisture for the new species that will create the undergrowth and, eventually, a forest of evergreen. Joined at the root within each grove, they grow in a related sisterhood, sharing a common fate of health or infestation, or vulnerability to falling in the wind.

Can we take off the blinders and recognize the echo of our own fragility in that of nature? Can we be reminded by the Sioux—whose moccasins have always been so finely beaded—that our feet were to be as beautiful as the earth we walk on? Or, in denial, are we headed more

for the fate of the Australian plant that must be burned in the ground fires of the outback before it can flower? We have known such consuming fires: Hiroshima, the Watts riots in Los Angeles, Exxon Valdez and the Gulf oil spills, of course 9/11, and that year, the day after Thanksgiving, revealing our fixation, a salesclerk trampled to death in the name of discounts at Long Island's Wal-Mart. Will we, how can we, meet the call?

Fueling urbanization, the proliferation of oil and gas "exploration," often speculation, creates anguish and shock waves in many communities. In Canada, Albertans are seized with the dilemma of the tar sands. Whether done by Exxon, Tecton, or Triton, drilling threatens the land itself. The odds seem overwhelming for the widowhood of the land. In April 2007, in the listing of threatened lands in a Wilderness Society's pamphlet I find that twelve out of the seventeen western public lands slated for oil and gas drilling over the next fifteen to twenty years are to be found in my own treasured public lands of the Rocky Mountain/Río Grande corridor.

Despair is dangerous, itself one of our dilemmas. I hold it at bay as I chant the place names, imagine the skies and the peaked or rolling beauty beneath: Clear Creek Fork Divide; Grand Mesa Slopes; HD Mountains Roadless Area; Roan Plateau Progress; Vermillion Basin; Bear Tooth Front; Bridger-Teton National Forest's Wyoming Range; Red Desert and its 270,000 acres of productive wildlife habitat; the Atlantic Rim; Upper Green River Valley; Rocky Mountain Front; Otero Mesa's 250,000 acres of grasslands, still threatened in 2011. Over one million acres of wilderness at risk in Colorado, New Mexico, Wyoming, and Montana. One hundred eighteen thousand new oil and gas wells bring industrial pressure more than close to home. Every project means habitat fragmentation for both humans and animals.

How can we deal with this? The work is long and political. Sometimes news of a possible, better understanding is encouraging. In April 2009, after eight years of efforts by hunters and other sportspeople, environmentalists, and a coalition of conservation groups, the US 10th Circuit Court of Appeals in Denver ruled in favor of the New Mexico Wilderness Alliance's case to protect a quarter million acres of Otero Mesa from oil and gas drilling. The case may go to the Supreme Court, but in the precedent-setting decision the three judges reasonably cited the need for "a very delicate balancing" for the aquifers and the one thousand wildlife species and two hundred migratory bird species inhabiting this rare Chihuahuan Desert grasslands, one of the "most endangered ecosystem types" in the country.

The Valle Vidal, a sometimes lush New Mexican mountain basin, and its abundant wildlife and cattle have been protected—it is worth the hopeful note—since the Valle Vidal Protection Act passed into law in 2006. So, yes, for a place of hope, gathering savings again, I fly home then drive the four hours with friends to this open wildland. Abundant elk and cattle roam freely in the 102,000 acres, but Comanche Creek, the four-foot-wide thread of water that runs through the valley, needs restoring to give back its original habitat to the cutthroat trout. I set to work with other volunteers from the Quivira Coalition, a conservationist group.

Close to the banks, three of us repeatedly unload heavy rocks from the saddlebags of Julie's patient horse. Kathy and Brad, a duo on horseback, will spend the long weekend riding through the meadows, keeping us and the other teams supplied with water, juices, and snacks. Group leaders pile posts of different heights we will need near each site. Dotting the landscape, the crews—young and older, awake, eager—are ready. By hand, carefully and together, every eighth of a mile or so along this river as wide as an acequia, we build small rock walls and hammer in the posts to hold them, making vanes to create healthy meanders for the current. Handling the rocks, even at moments fondling them, marveling at their shapes and the puzzle of piling them at just the right angles, we keep a steady rhythm.

One afternoon I switch teams for lighter work, harvesting willow saplings with four other women further down the valley. The crop is abundant, the pickup full. Returning to my first site, we spear narrow holes along the banks, tuck the saplings in the damp earth, pat them in tight. The graceful plants will stop erosion, freeing the creek of silt. It is soothingly and soberingly beautiful to see the water take its new course so quickly, freshened and, yes, already freed of silt and debris. Standing quietly in the vast mountain air, we watch its flow, our tenderness restored for this creek, this earth, this valley. Despair is dangerous in any universe. Repairing one habitat, working hard, feet wading in the shallow water, trailing through summer grasses in the company of good-humored voices, we hold it at bay. The last teams on a three-year project, we're proud too when the river engineer, Bill Zeedyk, confirms that with our work the plans have reached completion.

Mingling with the sounds of the work in the water and stories of the numerous bear sightings in the campsites, the name Galisteo has rung through our conversations. The Galisteo Basin, a breath-giving spread of wild and ranchland, home to major ruins of ancestral Pueblo dwellings and contemporary homes, is fifteen miles from Santa Fe. The oil and gas threat had moved even closer, as the New Mexico public learned the year before that sixty-five thousand acres of mineral rights had been acquired by Tecton Energy of Houston for oil and gas drilling there. Rumpled by wooded hillsides, the Galisteo's wide flattening bowl sits quietly open under the fathomless expanse of space we call the sky here. From their homes, often invisible in the landscape, inhabitants cherish the slight rainfall and relish, or warn about, the sightings of wildlife. From here deer, rabbits, pronghorn antelope, and mountain lions navigate the arroyos, streambeds, and wetlands—as well as current fencing and roads—along the long migration corridor stretching up through the Continental Divide into Colorado—the "spine of the continent" that eventually reaches the Canadian Rockies. For humans, the hours spent hiking in the dry, gentle warmth soothe the soul. Mingling with the anxiety at the news, there is still wonder at the petroglyph panels—war shields, serpents, snakes, birds, stars—etched in the black lava surface of a volcanic dike rising among the open grasslands. Incredulous voices ask, "Drilling? What now?"

Similar to the ranchers in Wyoming, Colorado, or Montana, whose hearts and land have been broken by oil and gas drilling, landowners in the Galisteo Basin have discovered that, due to provisions of the Federal Homestead Act of 1916, they do not often own the mineral rights under their land. Fortunately, the uproar in the region is immediate and palpable, the arguments adamant, the citizens galvanized, and their organizing effective. "It is not time to lose courage," insists my friend Louise, who is fighting the drilling. "We are born into this time; this is the job we have to do." The people's determination matters. Citing the risk to groundwater, the governor initially declared a six-month moratorium on the drilling, which was later extended, matching the county's year-long hold on the permit process. Resistance and knowledgeable concern among the people and their leaders made all the difference. A county ordinance setting new standards passed before the end of the year.

In this land of little rain held dear by its inhabitants, where drinking water itself needs protecting, 1.3 to 1.5 million gallons of water per exploration attempt would be used, multiplied by an average of seventeen attempts per well, according to Halliburton. As with current techniques in uranium mining proposed in the northwestern area of the state, water itself is used in the drilling, the pressure it brings to bear in the borehole "fracking"—as the industry terms it—the earth to find the oil. Toxic chemicals, in quantities exempt from restrictions of the Clean Water Act, are used, too. Recent, heart-wrenching cases of illness from water contamination in Colorado, Alabama, Ohio, and Pennsylvania have been documented, the individuals' stories told, confirming Santa Feans' fear of the "exploration" poisoning waterways or the aquifer itself.

Fracturing the earth
How careful we must be
How careless we are

When asked at a public meeting what the company would do if wells were poisoned in the Galisteo Basin, a Tecton partner, apparently ignorant of the desert ecology, asserted what citizens here consider—know to be— impossible: "We would find a way to replace the water." Some of the audience laughed in surprise. "How!? Where would it come from?" The painful grimace on a friend's face spoke as loud as her words. Knowing the already imposing demands on the Colorado and Río Grande rivers' water, serving seven states (which include the cities of Las Vegas and Los Angeles), I wondered, "Are they thinking of trucking water in from Canada?"

Well-founded too are the fears of death to livestock from the growth of noxious plants like halogeton growing in disturbed soil, due to the elimination of local vegetation. In conversation with two ranchers from the northern Rocky Mountain corridor, I didn't want to believe the sad truth of twenty sheep dead in one week on one ranch—one hundred over the period of a month on another. It was hard to resist the feeling of impending doom. News of the animals' deaths gave me an excruciating image of another (to use the scientists' term) "tipping point": these ranches' lush expanses of peak, pasture, and valley, contrasting with the areas where deer and sheep trails are cut through by roads, crisscrossed by trucks; meadows and sage plain bored with wells whose flaring streams of unused gases blacken the high-altitude air.

As the fight for the Galisteo Basin gathered momentum, I read that, further north in the Colorado Rockies, drilling activity had already begun in the pristine Piceance Basin—7,110 square miles of wild habitat. My heart sank at the seemingly unending struggle. The Colorado River runs through these high plateaus and deep valleys, its water used here too for drilling the thirty-three thousand wells planned for. Trucks now—eighteen-wheelers—roamed among the wells that have gouged the land, pipes for natural gas embedded below. The extravagantly handsome and threatened sage grouse, along with the largest herds in the continent of deer, elk, and the lithe pronghorn antelope, are losing ground to the wounding of the fabric of the wilderness that supported them.

Fracturing the earth
How careful we must be,
How careless we are,
How driven

There is a familiar echo of the 1970s as I read the words of a regional executive director of the National Wildlife Federation in the newspaper: "I think we're watching the end of the West." He too objects to "the post-stamp" areas we are creating, where remaining national parks and the ecosystems they protect will become isolated areas while the wider land is converted to industrial-scale productions, interrupting or eliminating life cycles. Human disturbance.

The tenacious beauty of the beloved empty spaces—the "nowhere" but our true somewhere of the West, the fur traders' "Wasteland"—has become for the real estate and energy industries the land of plenty for the few. With plenty of destruction, plenty of loss, plenty of long-term devastation of resources comes plenty of money for the government, developers, and corporations.

While the global financial crisis reigns, affecting the daily lives of millions of us, US and Canadian companies, according to the *Oil and Gas Journal*, continue exploration and exploitation of hundreds and thousands of acres in Texas, Indiana, Australia, Columbia, Libya, North Dakota, Papua New Guinea, Louisiana, and Mississippi. Along the Continental Divide down into the Galisteo Basin, companies are abandoning wells, moving their rigs to Wyoming, Texas, and other areas that are less regulated—not accepting easily that the rules of the game are changing—and hoping still to "explore" sometimes now for briny, deep aquifer water five thousand feet into the earth. Exploration which, they know, as a by-product might by chance produce oil. The impetus remains. So do the close to sixty-five hundred drilling permits accorded in Colorado alone in 2008. Greed can still govern over concern. What will we do with it? How can we get out of the snare?

Amid the restless wells and silenced homes across this continent, we can't help but see how much we have lost. A dream image comes to me of hundreds of us standing, disheveled, in a wide plain of disarray. Where is the place of soul and soil here? It's glaringly clear how we have yet to embrace ourselves as part of the natural world; vividly clear how little means we have given ourselves for enriching the spirit of our culture rather than our pockets; not yet clear how we will stay alive together, move along whole again with the land that remains, and refurbish our courage for a responsible society and an identity that means something in human terms.

If in North America we are acknowledging more openly that we have been sacrificing the ground we walk and drive on to money, we have yet to see if extravagant profits for individuals, corporations, and governments can bring back the honeybees to gardens or thousands of people to their homes. This money will not replace the clean air, water, or forest. The naturalist Robert Finch, speaking of a wetter habitat in *Louv's Last Child in the Woods*, reminds us, "[The landscape] may still be kept technically alive—with sewage treatment plants, compensatory wetlands, shellfish reseeding programs, lime treatments for acidified ponds, herbicides for . . . ponds, beach nourishment programs, fenced off bird sanctuaries, and designated 'green areas'—but it no longer moves or, if it does, it is not with a will of its own."

That moving "will" of the land's own is what creates the "love" Corbett speaks of. It is a fundamentally reciprocal relationship. Without it, the land's health, our health, the holy, the joy, the fun, and the future have gone out of it. But whoa! Despair again, that disabling enemy. Westerners do share a profound sense of the land as sacred in spite of the evidence that land greed remains lusty. Who will rein it in? People will. People have begun. Too few as yet, perhaps, but ordinary citizens, ranchers, city and county councilors, farmers, mayors, land-use lawyers, conservationists, water experts, and governors are involved. People reviving the land's health, hope, knowledge, and community spirit are working together and succeeding in protecting the land.

And the news comes, several months after the Comanche Creek restoration: the Galisteo Basin will not be drilled! Friends in Santa Fe send me news of a celebratory gathering. There will be room for the will of the land. Under the pressures of a tanking economy and the new county ordinance, Tecton ended its plans to drill. Another collective sigh of relief, even jubilance, could be heard. Yes, in this place of breath and beauty, let there be room for the will of the land and the animals. The groundbreaking ordinance makes small as well as large exploration and drilling companies responsible for a comprehensive plan for water resources and adequate infrastructure for safety and fire before the drilling begins. Most significantly, it renders the companies responsible for avoiding ecologically fragile areas and habitats, as well as insisting on the use of nontoxic chemicals the county identifies.

I learn another bit of good news about the same time. In the Piceance Basin, half the oil rigs have been silenced. Colorado too has adopted regulations more protective of habitat and wildlife. But the osprey and great blue heron there will fly over the wounds—seven thousand wells already drilled—that remain.

Attending a conference in 2007 entitled Liquid Assets, dealing with water issues in the Southwest, and again at an environmental conference of the forward-thinking Quivira Coalition in 2008, I hear calls for nongrowth policies in the Southwest. "Nongrowth," "slow-growth," "green-growth"—the vocabulary grows as efforts are made to reimagine our use of the resources, reimagine a less wasteful way of life. We human animals are predators. We need food. All of our possibilities, all of our solutions in a rural or urbanized world imply the use of land and water—and our care. The health of both demand our care. More reasonable models of development that hold the health of the land and communities in mind are being designed and proposed in New Mexico, Colorado, California, Oregon, and other states. Can they insure that the eternal strength of mountains will not be reduced to dust? Hope rises as we see more scientists, land-use specialists, county and municipal councilors, mayors, governors, and their populations struggling to find ways to protect aquifers and groundwater while respecting the necessity of open space for children and wildlife to roam.

On a humble scale but with a sense of common purpose, individuals in Albuquerque and Santa Fe, faced with possibilities of private wells drying up, retrofit their homes for water catchments and rain harvesting. Drip irrigation has already been increasingly used for twenty years. There is new interest in and motivation for solar home and water heating systems, as the technology improves over that used in the region since the 1960s. Like farmers in Quebec City who now have a permanent space for selling their produce, farmers in Northern New Mexico and their clients are enthusiastic about their permanent structure, the Santa Fe Farmers' Market Pavilion, housing their weekly market, abundant in healthy local food. Classes for training in green building and renewable energy, including solar energy training programs, at two of the state's community colleges have been full since their inception. Volunteers are numerous for projects helping to restore the watershed. Once drinking water can be protected, local voices support prioritizing water for local agriculture. This time is our chance, a still-urgent one, to turn the "tipping points" into turning points by opening the floodgates of our—the people's—steadfast will to do so.

On a larger scale, the industry's drive is tough competition. Suddenly, in the grip of rising food prices, fluctuating gas prices, and a plummeting stock market, it is clear to us that more radical changes are at hand, and necessary. Clear that we need a fundamental restructuring of our energy base. But how? In a television interview on *Meet the Press*, T. Boone Pickens, the elder and sometimes controversial oil magnate, stands by his commitment to wind energy as the "cornerstone" of his plan to develop cleaner energy within the United States over the next ten years. Wind turbines built on his own Texas ranch and up through the windy western corridor to North Dakota would bring electricity to 20 percent of US homes and businesses within ten years. Passionate, he knows firsthand, as do the environmentalists, "the simple truth that cheap and easy oil is gone" and the country is in dire need for renewable energy sources. Randy Udall, an authority in the Unites States on sustainable energy and peak oil, points out that prior to 1900 oil consumption did not outpace biomass fuels like wood. In the twentieth century, though, with our addiction to mobility and flight, there was an eightfold increase in our consumption. Two-thirds of the oil resource deep in the land is already gone.

Ted Turner, the media magnate who loves the land and owns more than a lot of it, with his fifteen ranches in Montana, New Mexico, and five other states, shares their conviction. Listening to his interview on *Meet the Press*, in which he describes a digital grid transferring electricity to the East Coast from locally developed solar and wind technologies based in resources in the West and Midwest, my mind boggles at the sheer physical vastness of the enterprise before our leaders and workers, not to mention the clarity of decision needed to set it in action. Knowledgeable partnerships, perhaps, with landowners, scientists, and government agents, coupled with private and government funding, could lead to the major changes that are necessary.

Elsewhere, visiting the exhibit Feeling the Heat: the Climate Challenge, at the Birch Aquarium at Scripps Institution of Oceanography near San Diego, a California friend and I are heartened and astounded by research showing that "the United States receives enough sunlight to meet its total electricity demand ten times over with solar power"—ten times over!—"and enough wind to generate up to three times the amount of energy produced from all sources in 2006." Although most of our production of solar panels made in New Mexico is currently exported to Germany, we can be confident for the future that, as with "clean" cars, the capacity exists.

And food, what will we do about our food? There are the farmers' markets, their popularity a motivating hope in spite of agriculturists' worries about demand outpacing supply. In cities from the rainy Northwest to Harlem, gardens and small-scale urban agriculture thrive here and there at the hands of the people in backyards, vacant lots, and on rooftops. Nevertheless, a successful, clear-minded Southern California farmer I met on a plane recently was agreeable to discussing the anxious fate of farming. He predicts, in spite of his family's four generations of success, that his large farm will suffer death to urban sprawl by the next generation and that, on a larger scale, by 2050, Americans will be eating exclusively imported food. Perhaps. Statisticians agree. Already in 2007, 80 percent of seafood and 45 percent of fresh fruit eaten by Americans was imported, according to the Center for Food Safety at the University of Georgia. Michael Doyle, the center's director, foresees that the United States will be a "net food-importing country within twenty years." This is sobering. These predictions (realistic, I understand) are juxtaposed in my mind with my experience on the ground meeting ranchers in Northern California and the Rocky Mountain /Río Grande corridor whose controlled grazing and monitoring of grasslands—one and the same with wildlands—counter old habits of wasteful overuse. Today they, too, must resist the encroachment of housing developments and off-road vehicles. Giving the landscape long months of rest after a week or two of intensive grazing, they allow the grasses to grow and the watershed to return to its self-willed well-being, producing healthy grass-fed cattle and hope—fragile, perhaps—for our food production.

Many of us are landless. What will the millions without gardens or soil do for fresh food, safe food? On a CBC radio interview, I hear the voice of another committed individual who has something different in mind and on the drawing boards. Dr. Dickson Despommier of Columbia University sees promise for urban food production in New York City's canyons: vertical farming, or agriculture in skyscrapers, a variation of greenhouse farming used in more northern countries like the Netherlands and Quebec. Freeing up much of the arable land in the Midwest to return to grassland or forest, avoiding genetically modified crops, and reducing waste and reliance on fossil fuels, the grains, vegetables, and fruits—even poultry and pork—raised in vertical farming would grow to health, thanks to hydroponics, solar, and wind technologies. Searching for these multistoried, urban farms on the Web, I am struck by the beauty of the architects' designs of the "Living Towers." They are jewels reminiscent of hanging gardens and of the Austrian painter/architect Hundertwasser's vibrant work.

Hundertwasser, who chose New Zealand as his adopted home, created a verdant world on canvas where raindrops are precious gold, all the while pioneering an ecological vision for everyday environments, building homes, churches, and apartment complexes using circular forms where the body is welcomed, rooftop gardens where food flourishes, and trees growing inside apartments, their branches adorning windows. A rich bouquet of life-work. Perhaps, like Dominique Mazeaud's ritual work in the Santa Fe River, this early ecologist's pioneering on a simple scale nourished our collective imagination.

And more science is available now. As I realize Despommier's vision is being taken seriously, it seems we are readier, more knowledgeable, and willing to harmonize our food production needs with nature's balance, enhancing the big city's beauty as much as the open range's. Wherever we are making the changes, the repair of the economy and the repair of the land need to go hand in hand. It is more than time. In the tone of our voices, the pleats of our brows, the catch in our throats, I can see we are aware; our words have become more than just new vocabulary, more than just media sound bites. We know now. Destructive human disturbance is not the only option. Yes, we are predators. We intervene out of necessity, but the wild and its human fabric, our human presence, can be more gently, more finely woven. It is an old lesson, that of treading lightly on the earth, if we will heed it.

Distracted, embroiled, boggled, and startled but not, I believe, totally defeated by overconsuming, Americans' corporate and middle-class values have been in a distressing moral crisis. But on every return to the places and people I know, I find a wealth of human kindness, generosity, tenacity, integrity and, yes, particularly in the dry and developing West, the hard, gritty, and heartening work of survival—all holding against the odds of fear, greed, and confusion.

Speaking of woodlands, sea, rain, soil, and grass in his 1973 exhibit catalogue, Hundertwasser encourages us, saying:

We live in PARADISE
but we do not know it.

Do not destroy
do not revolt
do not escape
just IMPROVE slowly . . .

As we move forward, thoughtfully working toward a prosperity that respects the earth's needs and ours upon it, so much still depends on those of us—individuals, families, communities—still living in this troubled world. If we care for this paradise in all its delicate exuberance, if we believe in and cherish the wells of courage, thought, and capacities within ourselves for restraint, creative innovation, careful intervention, deep impetus toward community, and the rhythms of more selfless attention to the earth, we will be able to sing along with Louis Armstrong, "What a wonderful world."

FIFTEEN

Land of Kinship and Innovation

If kinship is deep in our bones, the West is too, a land of innovation that de Tocqueville would admire. But there is still history to deal with. Government ownership of land has been at the root of much drama west of the Mississippi, a world that de Tocqueville did not know, having seen the United States only in its shaping in the 1830s. The philosopher Frenchman had no contact with what was then Mexico. Though he seems to have enjoyed the company of Native people east of Oklahoma, he surely did not witness the Pueblo Corn Dances or reach the two or three hand-built forts along the walls of the canyon south of Jemez Springs, a few miles from Jemez Pueblo, where courageous people defended themselves against the conquistadors in the 1600s. Neither would he have seen the grasslands of the "Wasteland" the vaqueros came to know late in the nineteenth century, or witnessed the American presence taking hold with the Indian Wars, as history calls them, or the development of public lands in the early twentieth century. Forts, dances in the plaza, cattle on the range, government institutions in the land, a full spectrum. Whether of colonization, celebration, or war, the land carries all the memories we give it.

Since the founding of reservations, national forests, and national parks, the challenge of public lands has plagued both the West's earliest and newest inhabitants. Washington's power, considered foreign then (and still now to many westerners), led to a new makeup of American society where, rather than the townships of the founding fathers being at the heart of Anglo-American development, government ownership of land became a major force in defining the country and its people's relationship with it. Still today, 50 percent of ranching and farming land in the American West is owned by the federal government, under the auspices of the Department of Interior. Once dispossessed of their original homelands, Native communities on their reservations have always had to negotiate their freedoms with the Bureau of Indian Affairs.

Unfortunately, in the early twentieth century there were also failings and contradictions in Teddy Roosevelt's vision. Rural Hispanic villages know how, by imposing restrictions on the traditional people living on the government-claimed land, the Bureau of Land Management did not recognize their subsistence needs for harvesting firewood, water, and materials for building their homes. Native communities know how the importance of hunting for physical and cultural survival has so often not been recognized, and still today, in New Mexico's high-altitude forests, the conflicts over grazing rights burden both the "feds" and the sheepherders, while battles wage over mineral rights between homeowners and oil and gas companies. Families like mine have more recently lost their old homesteading cabins in the Pecos Wilderness. One's home, one's belonging to the land, being entwined with institutional powers, while not new still remains a widespread dilemma in the West.

As a result, grim mistrust between ranchers and farmers and the federal agencies ingrained itself in our western culture. It is the main reason I had understood as a child that federal intervention in our western affairs was not welcome. Curiously, my mom's reserved appreciation and my dad's open admiration for the East Coast and founding fathers of the United States coexisted with the sentiment that the bureaucracy in Washington, DC, was not to be trusted. In turn, knowledgeable ranchers and farmers have gritted their teeth and bitten their tongues at simply not being understood or trusted by Washington—and by many conservationists—when they know they are responsible for our country's food supply, leading a difficult, demanding life motivated by the love of the wild and the animals. "Yet," a friend points out, "due to accepting necessary subsidies grudgingly handed down by the feds, they are accused by environmental purists of 'cowboy socialism.'" The respect they deserve is hard to gain. At the heart of our society's food dilemmas, these people were and remain the primary stewards of our land—mountains, pastures, watersheds, grasslands, arroyos, forests, meadows, peaks—beautiful lands held in intricate administrative and cultural relationships that have had difficulty fostering the true spirit of commons.

In today's complex web of relationships determining who claims ownership—private or corporate, state, federal, or conservationist—of water, grazing, and mineral rights, all the parties involved know the land needs care. Above and below its surface. It is changing under the country's hands and under global warming, with the changes themselves not always immediately measurable. The task, more unpredictable due to climate change, demands repeated adaptation. If westerners recognize that, historically, poorly informed livestock or logging policies and practices dried up grasslands and depleted watersheds, they know that the rampant paving to accommodate urban sprawl and its long reach—compounded by the proliferation of oil and gas wells—has even greater impact on the soil, atmosphere, and waters our lives depend on. The land in itself is disappearing. The time needed to remedy the situation is short. It is late already! Fortunately, in quitting denial, willing Americans are finding strength and effervescence in a more widespread environmental awareness. "It is vitally important for people to understand that knowledge makes the difference in how we approach everything we do," says one rancher. Vital, too, the awareness that managed grazing is a tool today for spurring and nudging land health.

The moral issue of caring for the land—the valid moral imperative that Native American cultures uphold and that Teddy Roosevelt did see—implies responsible stewardship willing to protect the ecosystems underlying the beauty of our habitats. Today on reservations, in research institutes, on farms and ranches, in national parks, universities, and federal agencies there are people capable of creating the discernment and tools for the contemporary adaptations we need. And they are talking to one another. The stewards face a tall order: restoring watersheds and health to both commercial and wild herds; innovating for energy development that limits drilling and mining but serves the regions and the country at large; and guaranteeing a future for agricultural families and food for the whole country. Our long-term thriving depends even more now on the better American "manners," as de Tocqueville said, "of cooperation and independence" (so different from extreme individualism) that allow human communities to function in and with the land. In his speeches, Roosevelt often mourned the destruction of wildlife species he witnessed in his day. Yes, the lives of the wild carry significant living knowledge. Our perishing is linked to theirs. We are listening again.

Where does one turn to find the West alive? It's true the land is under duress everywhere, but beyond the shock of our "transitional time," like its inhabitants of the 1980s that William Kittredge spoke for, we can look to the horizon. This time, we find a new gathering, a new bonding of westerners across traditional and political, even geographical divides. Among them is a deeply hopeful interest in working together. Through mutual respect, the sweat and thinking of it, the tears, joy, and sound knowledge of it, the sustained effort, new solutions are being found on and for the ground. A saving grace, our spirit, is back in place. A commonwealth has returned.

The most reliable stewards of the land are the people who maintain intimacy with it: the Native, Hispanic, or Anglo ranchers and farmers; the cowboys (often maligned or exotic cultures in the mainstream American view); the long-established in-holders; the permanent rangers; the longtime residents of mountain towns and villages; geologists; hydrologists; ecologists; as well as new residents who know their landscape. These are the people who see and know the balance and beauty of the land as other than strictly a commodity or property. On the ground, range management brings into focus the root bond: the earth and "the sharing of fate," as farmer, poet, and conservationist Wendell Berry puts it.

Fortunately, the new century has brought broader gatherings of stewards with direct experience of the land, creating community among various land-based people from diverse areas across the country. If there is hope to be found in spite of the subdividing, the industrial exploitation of the wild, and our ensuing disinheritance as Americans, it is that what the West means, what living with the land means—even if retreated into hearts and memory—stays alive or is easily revived in those same hearts and in the hands of the people.

Ernie's lithe figure, warm gaze, and keen, open intelligence had been easily recognizable among the participants at Mesa Verde National Park's one hundredth anniversary. After many years without occasion to meet, our reunion was heartfelt, our friendship full and alive. Our original bond of the land, acknowledged in the late 1980s among the ruins of Mesa Verde and Chaco Canyon, had led each of us to witnessing it on different paths: through his administering protection of the land and my years of teaching Native American literature in Montreal and my poetry performances. Knowing this friend's commitment and his gifts for bridging his own Hispanic culture with people of other cultures, it is to him I confide (even declare!) in the fall of 2006 that I have come home looking for hope.

And hope I find the following January, on his recommendation, attending the Quivira Coalition's annual conference held in Albuquerque, celebrating the group's tenth anniversary. Founded in 1997 by a rancher and an ex-Sierra Club activist, the Quivira sought to break the gridlock between ranchers and environmentalists in favor of common ground, a "third position" where, literally, among the "grass" and the "roots" peace could be made in the grazing wars. Amid the early wave of land-based collaboratives, the Coalition aptly named itself Quivira, a term marking the unknown territory that Spanish explorers had indicated on their maps.

Much is unknown about the future of the new land collaborations they and other forward-thinking groups foster in the West, but here, gathered with them in a hotel in Albuquerque, the territory for me is recognizable. The joy of being among so many like-minded people is palpable, and I am deeply moved in finding the comfort and pleasure of my own original culture. The atmosphere is reminiscent of the old days, those family years at the National Western Stock Show. But these are not the old days.

The garb here is familiar. The boots, cowboy hats, fine belt buckles, skirts, bandannas, and caps that these men and women wear are not signs of an outsider's style or an appropriated way of life, but marks of real people still working the land. These are boots that know the flank of a horse; hats that carry the dignity of long hours with the cattle; minds that know and study nutrition to ease the weaning for lambs, work with ecologists to reestablish riparian habitats during drought, and pay attention to the legislative and regulatory processes in their state and in Washington, DC. These men and women work with computers, sophisticated soil studies, and videos as much as corrals, horses, cattle, and water troughs. Among the clamor to save the land, their voices are most often quiet and steady, using knowledge of the heart, the experience of labor and wonder, science, administration, and poetry for the twenty-first-century West.

At lunch I speak to a ranching couple whose despair for their parched land and dying cattle had reached rock bottom during a prolonged drought. But among the hopeful, heartening stories, theirs tells how they were able, in only six years, to turn the situation around, their land now verdant and productive for them and their healthy livestock. Through the Coalition, they had found new practices focused on the specific needs of their land, which complemented the knowledge handed down in their family.

Back in the Marriot's auditorium, I listen to the story of the founding of the Malpai Borderlands Group that the Quaker Jim Corbett, one of its first visionaries, had been a part of. Countering the despair at seeing the deterioration of cattle ranching in the early 1990s, he, Drum Hadley, Bill

McDonald, and a few other ranchers pulled together to address their mutual concerns about the impact of subdividing on their "working wilderness," as their website states, "a 1,250-square-mile triangle of land draped over the Continental Divide where Arizona and New Mexico meet the Mexican States of Sonora and Chihuahua." Affirming to each other the possibility of trust over suspicion, of solutions grown on common ground, of the force of willpower in common purpose, the land owners began to work across political boundaries to protect their landscape, their threatened livelihoods, and a way of life they all loved. Through reaching out and establishing themselves as leaders in land management rather than clients of government agencies, a new path for partnership was opened. Since 1994 the group's ranchers—together with scientists, related public agencies, and private conservationists—manage, protect, and repair more than eight hundred thousand acres of unfragmented landscape where the spirit of the "untamed" can breathe, where the animals, plants, and one hundred human families that inhabit it can thrive for generations to come.

The circle of trust at the conference has a wide reach, bringing Amish farmers from that east beyond the shores of the Mississippi, ranchers from Wisconsin and California, French sheepherders from across the Atlantic, and the Masai from Kenya, each presenter carrying stories that continue to inform, delight, and reassure. The Masai sheepherder reveals the subtleties involved in the balance of his nomadic people's cohabitation with the lion, while a French sheepherder describes the important (typically French!) "palatability factor" for deciding grazing patterns on alpine slopes according to the sheeps' pleasure in the flavor of various grasses. I am enthralled as I sit immersed in these hopeful sharings of knowledge, enveloped with the beauty of a recognizably living twenty-first-century West. A PowerPoint presentation unfolds with images and stories of the intergenerational work on Sharon and Pat O'Toole's thriving and long-beloved ranch in Wyoming, with its meadows, healthy cattle and sheep, flowing rivers, and seasons of transhumance. Suddenly, though, there is an excruciating, heartbreaking contrast, a reality too. Drilling: thick black clouds of escaping gas floating above antelope and elk habitat on the O'Tooles' section of the Wyoming Red Desert's Atlantic Rim.

People of the New West—these individuals, families, and groups—embrace the challenge of moving ahead into unchartered territory to, in the words of Quivira's visionary founder Courtney White, "repair and maintain land health" in the West. Realizing that solutions to problems of the wild and working landscapes are both social and scientific, they invest in establishing the all-important trust among various parties for finding ecologically sound knowledge that allows innovation in restoration and ranching practices. In a more academic setting in Colorado, the Center of the American West's good-humored and well-loved historian Patricia Limerick acknowledges common ground too among historians, engineers, biologists, and religious studies scholars. In her lecture The Rise of Regret in the American West: How to Tell Meaningful Action from Pointless Wallowing, she says, "I have been fortunate to have had quite a number of experiences like this, of moving from opposition and bitterness transformed to friendship, and that is one element in itself of why I am drawn to the idea of the Healing of the West." Healing the West. People and land. Grounded in a movement imbued with the best of the Western Spirit, these relationships give shape, place, and voice to the deep need for effective collaboration among those still cherishing the land and the way of life it has offered.

The Quivira gathering gives me a sense of a second homecoming, having the same power of identity and community that the petroglyphs had given me so many years before. Moved deeply by a sense of at-home-ness with these people who, for the most part, don't yet know me, feeling the sadness of a life spent (at the time) so far from them and this land, I weep quietly off to the side during a break at the conference. Unexpect-edly, a beautiful, dark-haired woman responds with genuine warmth to, and quick recognition of, how real my emotion for the land is.

At first sight, her rooted spirit speaks in the proud but unobtrusive way she wears her turquoise jewelry and ample fiesta skirt, in the tradi-tion of women I knew growing up. Her calm stance and generosity are natural to the land and culture she comes from. Tuda Libby Crews is a dedicated rancher in northeastern New Mexico. Having restored her family ranch from four pastures to eighteen, she and her husband Jack reveal the extent of collaboration involved in healthy ranching in the twenty-first century. Tuda's life is a labor of love: smart love, tough love, kind love. Like many stewards of the land, she and Jack exemplify "the spiritual and practical calling" that Jim Corbett spoke of, embracing its full complexity in the modern rancher's world.

Tuda knows gladness
Cattle and cattle pens
Cookie dough and cutters
(She has a way with dough!) and grandchildren
Cowboy poets and their videos
People, their fears and possibilities,
And now birds in blossoming bushes
and pride in the more than 170 plant species on her land.

On this day in late August, over a year after our first meeting, Tuda is on the road, passing through Santa Fe after having brought her ten-year-old granddaughter Bella to the airport in Albuquerque. Hungry for more understanding of how she and her husband Jack have created a healthy, holistic ranch, invigorated at the thought of our meeting, I am grateful she has taken the time to get together. We show up within a minute of each other in the St. Francis Hotel's pleasant lobby. Quickly engaged in conversation, we find a quiet corner table and order: duck salad and lemon shrimp, iced tea for both.

The Ute Creek Cattle Company is in remote and rugged country in Bueyeros, New Mexico, among the plains and mesas of Harding County. The story of the land where Tuda's great-great-grandfather and his brothers homesteaded in the 1800s would be familiar to many westerners. When the Dust Bowl intervened, its hardships obliged the family to lease out the land. Fortunately, Tuda's parents were eventually able to reclaim it, creating a home for her and her four siblings. Studies and marriage took the young woman to Arizona, where she met Jack, and then to Wyoming, his home state, where the happy couple in turn raised their daughter Libby and son Ted. But, Tuda acknowledges, "my heart never left New Mexico." She and her husband shared the determination to return one day to the ranch. In Wyoming, their closest friends were ranchers, so they were "never far away from the cattle industry." Range education, technology, cattle genetics were all part of the interests, even research, they maintained as they brought up their children.

As we began our conversation at the St. Francis, it was the grandchildren I asked about first. "What's most valuable in what you want to pass on when the children spend time on the ranch?" "It's so deeply important to me, instilling a connectedness," she says, her voice warm.

But when I ask, "How?" she responds, "Who is it that said, 'To know is to love'?"

Knowing the beauty, the hard work, the respect and reciprocal rapport with the animals, with seven generations of the place in her blood, she hands its spirit and way of life down to her three grandchildren—Bella and the twin boys, Bennet and Seth. They and their parents are part of the daily life when they come to the ranch, involved in activities appropriate to their ages. This grandmother's main concern, though, is to give the young ones "a great time" with no stress in their growing rapport with the ranch, learning the nurturing of the land and building loving ties with it. For them and their children's children.

Piling the family into their 1958 Willys Jeep for a ride out to picnics under "The Big Tree" has created a ritual important to bonding with their beloved empty spaces. As Tuda evokes the contented gathering under the tree and the playful delight with the kids, I imagine easily what it will mean to them, knowing how deeply my family's cabin has mattered in my own love of the land. These children will have something in common too with people from Botswana whose stories have shown me the importance of a village's "big trees" as gathering places for solving knotty problems, bringing reconciliation, or simply indulging in companionable gossip. In Bueyeros, horses too will play a role. While the twin boys, younger and riding less frequently, might be playing with two goats, Bella's grandfather is teaching her to ride and maneuver when they are among the cattle. Jack's words clarify the relationship with their adult children, Ted and Libby, and the grandchildren. "Unfortunately, they aren't right here with us on the place, but they sure are part of the operation. They're in mind on everything we do and every decision we make."

Tuda and Jack themselves returned to the ranch in 2001 knowing a big job lay ahead, especially that of keeping the work in line with their holistic management values. Over the years, the land had suffered from lack of care and all-too-frequent New Mexico drought. Like the Santa Fe River that Dominique Mazeaud had adopted, the ranch was in need of attention, the creek bed dry, the grasses spent, the herd half wild. The couple was well prepared and convinced that the family land could be sustainable—and would be passed on. A challenge.

In less than eight years, Jack and Tuda have brought the land from four pastures to eighteen through a rotation grazing system, producing three hundred head of vibrantly healthy cattle who meet the highest criteria for quality natural meat. They have restored the riparian habitats and watershed so sufficiently that the dreamed-for miracle of Ute Creek flowing all year long is a reality, a feat in this land of little rain. To my comment that their return late in life to this hard and risky ranch work is courageous, Tuda brightens, chuckling, "I just don't have a lick of sense." Then in a more settled tone she adds, "I think we came home with a fresh perspective. We looked at everything with fresh eyes. You have to be confident, and bold." "Bold," I say, "I like that," and two bold women laugh together, pause to admire their meal, then move to discussing the ranch again.

"I do an awful lot of cooking at the ranch," Tuda explains. Like the rancher woman she is, she cooks delicious and abundant food, knowing its essential power to sustain and bond those who partake: neighbors, family, friends, the workmen, and professional cowboys they hire. "Food is love made visible," her mother would say.

Food from the land. Food for the people. Healthy grasslands, healthy cattle, and healthy food are at the heart of this rancher couple's common vision. No antibiotics or hormones, but rather low-stress handling conditions for the animals, close monitoring of the grasses' growth, and controlled grazing help guarantee its implementation. A rancher's attention to his livestock making all the difference, Tuda credits Jack's kind calm and expertise in the animals' good care and handling, down to the choice of disposition as criteria for sire selection. The weakened herd they had inherited has been improved by cross-breeding, coaxed through genetics and care from a wild to a hand-fed disposition. "We love our cattle!" she exclaims, flushed, having told me of the working pens, built in the round, where cattle are not injured, where the design respects the animals' natural inclination to movement and their propensity to keep looking for an exit. In the pens, the two handlers—Jack and Tuda alone—need only assist, monitoring the animals' flow.

Enthusiastic in her realization that "short grass prairie and birds go together," Tuda revels in the joy of their twenty-three-acre wild bird sanctuary as it matures. Twenty-five tall trees have been planted, flowering bushes and currants attract diverse bird species, and both birds and wildlife quench their thirst at a wildlife water guzzler. A habitat for them and a sight for sore human eyes and dreary hearts, this restored "oasis" may eventually serve too the economic sustainability of the ranch as part of their vision of nature-tourism. "Ranchers sometimes are very tradi-tional in their operations, but," she says, "if your goal is to be sustainable, you have to be open to new ways."

Knowing she is working within a different model of ranching, "the bird lady" takes the guffawing others might do at her less-than-usual initiatives with great good humor. Planting several trees in the corrals so she and Jack could work in the shade had at first raised a few eyebrows. But low-stress conditions for the handlers are important too! On this range, there are escape ladders in stock tanks for frogs, birds, and other wildlife. When Tuda mentions they painted the cattle pens turquoise— her favorite New Mexico color—I can't help but see how much pleasure and delight are integral to the hard work on this ranch.

And what else does this "active environmentalist," as Tuda calls herself, do? Resourceful, like any self-respecting rancher, for, say, repairing her truck's windshield in a driving rainstorm, she is resourceful too for writing grants—necessary, cost-sharing grants for funding to improve the health of the range and grants for maintaining the region's history and culture. She has brought neighbors together in a community foundation called Area del Campo, where together they aim to preserve their quality of life and promote sustainable agriculture, as well as their culture. In nearby Mosquero, with members of the community in mind whose families or friends might need a place to stay during weddings or other events in their remote region, she has restored the rectory, a building that embodies memories for the local population. Busy "paying labor and materials, picking up windows, or driving to Amarillo, Santa Fe, or Albuquerque to get supplies, the renovation has been," she admits, "a lot to think about." Another building on the ranch taking on new life, soon to become the first museum in Harding County, is the post office that she remembers driving to at age ten. Her story brings back my own thrill behind the wheel at age twelve, thanks to a free-thinking uncle.

We finish our tea, the bill comes, the sun sets
more clients enliven the hotel.
We linger, we ponder, we laugh
I listen,
we hear each other's reverence for the land.

Tuda's gaze grows darker as she thinks, then speaks of the disconnect between rural life and urban people who don't know "what it takes to grow vegetables or beef, and how to get milk to the grocery store." Like Estevan Arellano, she is more than aware that "a lot of people don't have querencia," don't know the land, don't have the informed affection for it. Undaunted, she remains, though, "proud of what we've done." Wallowing in regret is foreign to this woman's mind. Open to possibilities but realistically concerned about the impact of fuel prices in our "nation of transportation," she looks into the future, knowing that ranching and ranchers, particularly those in remote areas, will suffer in the shrinking economy. Drought remains a risk and a plague. Last spring she and Jack were obliged to sell their calves early. "We were out of grass," she states simply. Undissuaded, knowing the vagaries of their chosen field, they believe in a ranching future for their family.

Seeing geographical isolation as a plus, Tuda is convinced that ranch tourism offers additional possibilities for survival. She has thoughts too for educational projects—an institute, perhaps a live-in situation for young students learning the way of life in situ on ranches in Harding County. Carefully managed clean energy development—critical for the country as a whole—may aid this region as well. "I just believe it can be done," she states, as she tells me of her family company entering into an agreement with another family-owned company that shares environmentally sensitive attitudes. North of Bueyeros on the Colorado border, the new Cimarron I Solar Project is capable of producing solar energy for nine thousand rural homes in Colorado, New Mexico, Nebraska, and Wyoming.

Like their more southern neighbors of the Malpai Borderlands Group and colleagues in the California Rangeland Conservation Coalition, she and Jack know the necessity of collaboration for sustainability. First, of course, at the ranch there are the workers and professional cowboys they hire. "To earn their respect and affection is very important to me," Tuda states, acknowledging "a familial connection" with them—a bond as necessary as the actual valuable work they do in branding or delivering calves.

Other levels of relationships don't necessarily have the "familial connection" of daily labor on the ranch, but Tuda feels they have been very fortunate to partner with so many trustworthy people in a long list of programs that are essential to their holistic range management. With their realistic, heartfelt commitment, she and Jack act, negotiate, and succeed with their vital and various partners: the Natural Resource Conservation Service; the New Mexico Water Trust Board; the National Wild Turkey Federation; the US Fish and Wildlife New Mexico Partners Program; New Mexico Historic Preservation Alliance; Oregon-based Country Natural Beef Co-op; Nebraska's Profit Maker Genetic Alliance, where Jack leases the bulls to improve their herd; and Texas A&M University, which gives assistance for monitoring the grasslands. Private consultants too are partners—notably, Kirk Gadzie in holistic management and Bill Zeedyk in water engineering.

Tuda is particularly proud of Harding County's Mosquero Municipal Schools being one of four selected in New Mexico to be among the Microsoft U.S. Partners in Learning program. "If you don't take action, nothing happens!" she exclaims, exuberant. "You gotta move!" Consciously working with others to turn around the decline of rural America, Tuda concludes with her goal that cooperation must lead to a win-win situation for all involved. "The partnerships must engage in working toward a shared goal." Similar to diverse bird species and short grass prairie, interdependence is the only way to stay alive and flourish.

As we walk out into the evening, I entertain thoughts of good food and dancing at their ranch tour later in the summer. Ah, dancing and a generous spread of food. Tuda and Jack Crews, their family, neighbors, and friends know the value of food and gladness. From the street I wave, watching her turn off my road in her husband's dark, blood-red pickup truck, feeling humbled by the generosity of their endeavors. Have I ever met a mover and a shaker as genuine and unaffected as Tuda, so straight-forward? Perhaps, but rarely. Then again, I have known that the open range breeds such frankness and gladness in souls imbued with knowl-edge of the soil, its earth, and the necessity of cooperation to sustain life.

Tuda and Jack are among the many innovative ranchers in the loosely knit but well bonded New Ranch Network, where they, along with farmers, scientists, consultants, conservationists, educators, and volunteers of the Pueblo, Navajo, Hispanic, and Anglo cultures—joined by various individ-uals from other lands such as Rhodesia and Australia—are active together in restoring the land in the West. They suffer the fluctuations and tribula-tions of the economy but work well beyond the despair and nostalgia of loss. Mindful too of the widespread disregard for rural life in America, these people do, as Patty Limerick encourages, use regret as "an absolute fountain of renewable energy," knowing, along with her and others involved in the healing of the West, that "the key is to make sure that this energy, once generated, has an outlet and a project and a route to action," be it with conservation easements, seed planting, brush control, road restoration, water harvesting technics, or training youth as media entrepreneurs.

Adopting an ethic based first on the belief that people and the land go together, these progressive landowners' approach to careful and knowledgeable management supports finding solutions to the economic concerns inherent in working the land and to diversifying ranch operations for future revenues. And the recent scientific knowledge for rehabilitation they employ leads to focusing on site-specific solutions, well-adapted practices in tune with their own local ecosystems. This has the additional advantage of keeping people humble and respectful, not only of one another's understanding but of the actual needs of a river, a meadow, a pasture, a prairie, a field, a mountain slope.

A habitat. In the habitat, fear of the feds—and very often environmentalists and conservationists—remains. Some families do have reason to fear that, due to certain federal regulations, a ranch would be shut down if an endangered species is found on it. One among them insists, "A rancher should be commended, not shut down, for nurturing a landscape that an endangered species can live and thrive in!"

Committed to the belief that common ground and common purpose can be found and that working landscapes can still thrive, these stewards do us proud, bringing essential nourishment for hearts and memory that can revive our hope and motivation for cherishing the earth, contributing in a major way to establishing mechanisms and a context for trust among the various groups that must—this is essential—work together. In our often divisive market society, bringing people beyond polarization to find one another in mutual respect is perhaps the Quivira Coalition's Malpai Borderlands Group's and Center of the American West's most valuable innovation. They work in the gap *between*, a cultural space where the gift of commonwealth is possible; a space that becomes the new place of meeting and solutions. Using a more scientific metaphor, rancher and teacher Julie Sullivan compares it to an "ecotone: the meeting of two or more ecological systems or habitats." She adds, "Possibilities increase in the overlap between systems: more diversity of species trying new relationships with each other."

This West. Made up of intricate relationships, yes, in beautiful lands, but fostering now healthier land, watersheds, children, animals, birds, and—for our future—safe food sources. In this motivating collaboration there is a jubilant spirit among the gatherings of stewards. In spite of the necessary skepticism and the pain of facing weighty, difficult odds, the resilience and adaptability of the people are still alive. The proof is in the pudding: the ranches and farms that have been resurrected; rivers restored; grasses, pastures, and species revitalized; the landscape with its cattle, horses, eagle, heron, osprey, and elk breathing again with a will of its own. The innovativeness, the manners of cooperation and independence of those de Tocqueville admired are vigorous and active among these people, capable of giving necessary cohesiveness to human survival and celebration on and with the earth.

Making sense of its history and the whole of the land, the kinship of a viable culture is at work as westerners remember and restore productive lands along with the chemistry of beauty and the joy of solitary or shared wonder in the silence of the land.

SIXTEEN

Reweaving the Tapestry of Hope

What is savage is in the deepest sense gentle and what is wild is kind," Jay Griffiths reminds us in *Elemental Journey*, the chronicle of her experiences in Australia. She and I have this knowledge in common. I recognize a soul sister in her words and the echo of the gentle culture I grew up in. Her Aboriginal friends taught her that "land needs to be visited, like a relative."

Born to joy, I revisit Taos. The road north from Santa Fe hugs the shore of the Río Grande, flowing mightily today as a new ration of snowstorm melts into the narrow canyon of the mountain stream that it is. I stop the car. Stand, free. Gaze into the Río's waters recomposed by the light. Frothy waves push through its depths, splash or trickle into the sand on the shore, tumble over rocks where pools reflect fragments of the forest and our own image. From this shore it is not about our narcissistic infatuation with ourselves but love of the beauty we belong to, the reminder we are whole with the world.

Lungs and soul refreshed, I drive out of the canyon. The road rises pleasantly, winding through familiar hills, then bends sharply, unfolding the sudden view. High in the Sangre de Cristo Range, Taos Mountain itself stands over the wide blue-green plain cut through by the Río Grande

Gorge. Always dramatic. Sheer beauty. You can recognize *prana* here: the land breathes with its sacred force, nourishes and strengthens body and heart. Stopping again, taking in this sight for sore souls, I'm aware of Blue Lake tucked high above in the wooded slopes, forever a sacred site for the Pueblo and, by association, many other inhabitants of New Mexico.

Later, the graceful slopes of Taos Mountain rise to my right, or behind me, or above me, as I meander and relax for a weekend in this old place that sealed my love of New Mexico at age ten. On Taos Plaza a historic marker now indicates the town was founded by the Spanish in 1615 and that the Pueblo people are believed to have inhabited the site as early as 1354. It's interesting to be reminded of the dates. But no Indian elders hang out here greeting the sun and children as they did earlier in my own personal history. The Pueblo and the town are very separate entities now. The plaza itself and the road around it are paved. Clean. No dust. Not exactly new, of course, but newer. The five-and-dime store I remembered on one corner is now a candle store; the grocery stores, of course, are gone. In a more recently built side plaza, the requisite T-shirts abound. At the El Mercado, the general and hardware store I had so enjoyed as a ten-year-old, Florence—an employee who has been with the owners, the McCarthys, over the last fifty years—shares memories and explains how the merchandise has adapted to the loss of business engendered by Walmart. The ribbons and textiles of my childhood experience are gone; and the building has become more like a variety store appropriate for today's clientele.

The wild has seemingly been swept away from the center of town, but it is tenacious and still close by. Memories of it and the amazing personalities of the town's adventuresome survivors and artists of the 1930s are protected in the rooms and paintings at the funky old Hotel La Fonda de Taos. I savor being here, wandering through the streets and little side plazas, talking to gallery and shop owners. The natural ease in our exchanges about their creative products, the gorge not far, or the Pueblo's dances confirm I am not alone in my awareness of the mountain's kind presence not far from us. Enjoying a copious afternoon tea, a friend up from Albuquerque notes that here the shop owners and personnel greet you like family wherever you go.

Kit Carson's actual home, now a small museum, is accessible on the street named after him. I had always avoided going into the home of this man who embodied betrayal of the Navajo among his many roles in the complex history of our West. Here the unforgiveable is humanized through the welcome given by his great-granddaughters and their moving stories of the great love between Carson and his wife as well as through realistic, riveting images of the shabby Taos and Santa Fe of his time. I'm told too that it was especially his interest in "different countries" that led him to leave Missouri and strike out on the Old Santa Fe Trail at age sixteen, heading west for the then Mexican provinces of Colorado and New Mexico, where life as a multilingual mountain man awaited him. I can't help but note a common curiosity with this frontiersman, having been drawn myself to a "different country" (but in the opposite direction!) as a young woman. I take in this man's history, with its tragic lessons of human contradiction, and move on. Next door on Kit Carson Road are the rows of worn and wrinkled leather shelved in the unpretentious secondhand cowboy boot shop, telling of the rigors of living with mud, dust, dung, and snow. The illusion of living without the memories of land here is not possible.

Survival in drought and snow along with traditional hospitality bond the community at Doc Martin's, the big but cozy bar lounge at the Taos

Inn, where the whole town turns out on Saturday night. Kids, parents, cowboys, Indians, ranchers, locals, and tourists gather and listen to my musician friend Joe West of Grandmother West's clan, who, I am delighted to find, is up from Santa Fe to play with his band. In spite of "rural sprawl" (exurban development), Taoseño's worry about—and rightly so—in spite of the commercialization, the T-shirts and growing sophistication of the designer shops I tap in on this evening to the old spirit of New Mexico. The place is as comfortable as a living room: couches harbor little families or lovers, friends cluster around others on stools. I relax into a spot a couple has offered me by the window, at ease with the neighborliness, the easy camaraderie among the people, rich or poor, rural or townsfolk. Scruffy or spiffy, laid back or pulled together, dropping by or settling in for the duration of the show, the crowd makes room for everyone. Smiles and hands welcome you in like, yes, a relative.

After the festivities, I stay overnight at the McCarthys' comfortable and cheerful bed and breakfast, graciously adorned with their family's Taos and New Mexico heirlooms and art. The pottery and paintings here are resonant with their roots and traditions in the community. Although the McCarthys themselves are away for the weekend, the welcome among the staff—so much a part of their family—is still kind, generous these many years later.

Before heading back to Santa Fe, I muse on that "different country" I had immigrated to. When I had first arrived in Quebec, there were species that meant "Canada" to me—the wild cod on Newfoundland's banks, now gone, and the migrating monarch butterfly, threatened now in its southern home. Times have changed for them too. Will there be "forever less beauty" in our times, as Wendell Berry's poem feared? The beauty we can breathe, that gives balance to our health and lives? I don't

know. But what I find again today in Taos is joy in the wild—understood among its people, respected, cared for, and shared. Like filaments in a spider's web, recent memories of Taos Pueblo's Christmas bonfires with the elders' soothing, enlivening chants in the star- and fire-lit village weave through my thoughts.

There is the most beautiful of empty land here. It breathes in the vast air, a bond, a peace, among us. It is language of the animals and birds, shelter for the deer, it is a path of migration for people, tracking territory for the mountain lion, song for the hunters and spiritual leaders. Talkative and alive, the Aborigines might say, it is the larger soul.

During the last summer of writing this book, Kathy and Brad, generous Santa Fe friends, offer me shelter in their comfortable home, alight with hummingbirds, wild columbine, the smell of juniper, and admirably tall grama grass—a place where I share in the chores with the horses and dogs and where, tonight, after the sun setting and the light pink fading to soft gray, you can sit in the cool air listening to the evening settle in the trees, the very slightest of crinkling sounds.

Hearing the almost imperceptible shift of day in the trees, I'm reminded of Bernie Krause—scientist, musician, composer—who, as a young man, lay in a cornfield one night and heard the corn growing. Eminent now in the field of acoustic ecology (along with Gordon Hempton and others), he has recorded since the 1960s the sounds of living things—cottonwood trees, meadows, frogs—offering through a new science valuable and enlivening insight into habitat.

A morning interview program on CBC Radio had brought this man's story to my kitchen with what I thought was original music. Immediately enthralled, I stopped cooking to listen. Early in Krause's career, while he and other scientists were recording bats in a national park, they kept hearing a curious background drumming on each recording. Eventually, they realized that the cottonwood tree nearby was producing the rhythmic sound. Its cells were exploding to maintain proper pressure during drought conditions, creating a tight ring in the tree (this wonderful, dry percussion I had heard) and a new habitat for insects that were drawn to it. Adjusting their instruments to capture these peculiar frequencies not available to the human ear, the first recordings of a compelling new science, acoustic ecology, were born. Musical-sounding terms were coined: "biophony" to express the collective tapestry of nonhuman sounds animals create, "geophony" for nonbiological natural sounds, and, yes, "anthrophony" for our human noise introduced into the soundscape.

In the same interview in 2003, Krause reported that thirty years before it had taken fifteen hours to record one hour of nature sound— "pure," alive without the intervention of human noise. One such hour was now taking two thousands hours to record, due to increased noise resulting from human activity. This is immensely revealing of the rapid change in the subtle fabric of our physical environment. As well, one third of thirty-five thousand recordings the scientist had archived in those thirty years were sounds of living things now extinct. Yes, we know, don't we? Our habitat is under duress. In urban centers like Montreal and Denver, doctors tell me that more and more people suffer from ongoing tinnitus, a neurological disorder of the inner ear, due to the plethora of noise and our busy "beeps" in the city. Clearly our nervous systems are under duress. Well, we have heard it before.

Or have we? Have we listened? Nonhuman splendor and beauty has much to teach us about an order and purpose in the wild that perhaps we too yearn for, or grieve over. If we listen to Krause's description of the importance of a frog chorus in his article "The Biological Effects of Noise on Wildlife," we can begin to understand better the significance of the erosion of habitat.

> Many types of frogs and insects vocalize together in a given habitat so that no one individual stands out among the many. This chorus creates a protectively expansive audio performance inhibiting predators from locating any single place from which sound emanates. . . . However, when the coherent patterns are upset by the sound of a jet plane as it flies within range of the pond, the special frog biophony is broken. In an attempt to reestablish the unified rhythm and chorus, individual frogs momentarily stand out giving predators like coyotes or owls perfect opportunities to snag a meal.

Vigorous protective patterns made by individual voices that vocalize "very much," the biophonist says, "like instruments in an orchestra." This chorus gives us a cohesive metaphor for culture—and reminds me of family. In 1997, a few years after the National Park had taken it over, I entered the Bamber cabin with the keys I wasn't supposed to have. I felt sorrow seeing it neglected, one end of the roof sinking, but the key worked. As I opened the door, the space came alive with our voices, lilting patterns, a precise, recognizable gathering of our sounds, our words, clear, harmonized, a sheath of music emanating from this sacred ground. "*Tusarnituq*," the Inuit might say, "beautiful sounds." Carried in the land itself.

As scientists continue to study bioacoustics in Yellowstone, Glacier Bay, Sequoia, and other national parks, they bring us closer to understanding the subtleties of habitat health, the whole that is necessary to protect life. Distinguishing between natural and human-made sound, and their effects, brings groundbreaking insight for understanding stress levels in very specific ecosystems for the plants and the animals.

The noise of jet planes overhead or other clatter from the cars and conveniences so much a part of our American way of consumption may appear to us to be less destructive than the Indian Wars, but the effects of our lifestyle are pervasive and often painfully acute for the natural balance on which people and animals depend. Acoustic ecology's findings recall Chief Sealth's (Chief Seattle's) warning, "the forests are full of the stench of men," although now it is the sounds of our creations. Do we feel his words resonate in our lives at this juncture of history?

In spite of the rapid increase of disrupting noise in even our most protected wild areas, in spite of the fact that time is not on our side, there is real hope that, if we pay attention to the soundscape, these studies can influence at least park management, if not immediately our urban developments. Certainly they point the way for human animals' health as well. Scientific understanding can help balance our habitat and reknit the fabric of our culture. Beyond consumerism, science serves necessary heart knowledge, both inciting our wonder and contributing to our capacity to create a new balance between ourselves and the earth's own. The frogs' chorus protects the group; the interrelated trees in the aspen grove stand tall or fall together. Our new insights can allow us to be humble enough to learn, as our ancestors did, from the other wondrous—and much more numerous—species on the earth. In wilderness is the preservation of our habitat too. Perhaps we can be inspired by it to look again at a much needed resheltering of our own communities.

At times I have wondered, "Are we—is our own viable connectedness as humans, as Americans, in shreds?" There is sometimes reason to fear. America the beautiful is an amazing paradox—in the extreme. Electing a black man as president shows signs we are healing from our profoundly scarring history of racism, but our twenty-first-century racial profiling has been devastating, even deadly, for many of our citizens. We must be vigilant.

In our consumer society where everything has its dollar value visible in the media and in our conversations, we easily become self-conscious as image-ridden people. We are not free from existing as commodities to one another. This can wear at the kinship among us and keep us distanced from our own nature and the natural world. We are beginning to accept the evidence that overt greed has undermined our cultural understanding of who we are, affecting the fabric and membrane holding together our own and the world's communities.

Through years of reflecting on my own difference in the two cultures I have lived in, I've understood more clearly that the drive our American Revolution gave us, coupled with our own original invention of access to goods for the many as an ideal, has led us to being, at heart, a materialistic people. We are more violent too than what we often choose to recognize, but also characteristically and genuinely open, generous, and more willing to accept change on our soil than the deeply embedded class structures of European heritage could allow on theirs. We are a strong people, a paradoxical people. One of our strengths is that we are willing to recognize our failings. Above all we are resilient in our capacity to revive, bond, cooperate, and believe in recreating a future. Convinced that this is a mark of our particular freedom on this continent, I have held on to my faith in my people. In spite of my culture shock in the

1990s, my experience in the West reassures me that we are capable of reshaping a culture that makes room for knowledge, action, tenderness, and respect for our own and other species and for our earthly places— with whatever ration of their own intactness is left.

As we imagine how to find who we are again, for ourselves and the world, we have much to be accountable for as a democratic people. A weighty responsibility. Our diversity is our strength, and debate in this wide country is at the heart of our democracy. Paradoxically, though, I worry that a weakness abides in our culture where our more thoughtful opinions about our country and its doings are often not discussed due to the pressure of—and our comfort in—conformity. Scientific evidence, particularly related to climate change, with its energy implications, has clearly been dismissed or ignored more recently in our history. Whether due to fear or our willingness to be ignorant, it is a profound, long-reaching problem.

Nearly two centuries ago, de Tocqueville described how this could hinder us. "In America," he wrote, "the majority possesses a power which is physical and moral at the same time. It acts upon the will as well as upon the actions of men, and it represses not only all contest, but all controversy." This, he seems to argue, is why he knows "no country in which there is so little true independence of mind and freedom of discussion as in America . . . freedom of opinion does not exist in America." Strong words! We might bridle at them. Or perhaps we do recognize this in ourselves. In the twentieth century, Theodore Roosevelt seemed to be concerned as well when he stated, as documented by the Theodore Roosevelt Association, "A typical vice of American politics is the avoidance of saying anything real on real issues."

Seen from the West I grew up in, the "majority" I learned about in school was rooted in a white, eastern-seaboard culture, which I later learned was often termed the "dominant" culture. But Catholic, Protestant, Italian, Irish, English, African American, Hispanic, middle class and poor, old-timers and immigrants, we felt comfortable with our diversity and we valued tolerance. Already being part a big family, I knew that was necessary. Then, too, there was belonging in the neighborhood, the schoolyard where we played marbles together, the mountains, and the high school across town, where my first love was handsome and swarthy. In a city that had not yet outgrown its resources, difference in my Denver childhood was less threatening then than it is now. That cohesive, albeit sometimes complacent, culture lies far behind us, having given way to much more complex relationships. Not only in cities but in some rural communities, a sense of refuge is hard to find. We are confronted from inside our skin and our beliefs. Democracy is a sensitive process!

Out of her pain and the exuberance of her womanly hope, the dark-haired woman I once was, journaling in her Santa Fe studio, expressed concern about difference in the 1990s.

> We are not truly tolerant yet, she wrote. As we come to seeing and accepting difference, she thought, we must be aware of the danger of conflict but root ourselves deeply in our desire to be alive, to love, to exult as a species, to come from a committed place of listening, of cherishing the touch of our skins, the rush of our blood, the exuberance of our sexes, the resonance of our voices, those of others and the hushed or thundering sounds of the land. Listening. We have to find ways to heal beyond the suspicions and puritanical strains the pilgrims brought to this continent and recognize our own natural state—even brain research is showing we are wired for joy, wired for caring for each other! Can we reclaim

a tenderness for the body and for being bodies together in the natural pleasures of breathing, wading in water, walking, laboring up the mountain trail? Can we really listen to the peace and pain resonant in so many human voices?

A tall order. But she was right. Difference is about being human.

In his campaign, now President Obama tapped into our founders' ideal of tolerance and encouraged again a culture of dialogue like the one that forged the strength of the first townships in New England, nourishing again our resilience as a free people. Unfortunately, obstructionism on the political scene and anxiety over the recession risk blocking our dialogue with anger and confusion. But sometimes we may need to be bigger than our leaders. Close to one hundred and fifty years after de Tocqueville's *Democracy in America* was published, I came across Adrienne Rich's thoughtful reminder in her work, *On Lies, Secrets, and Silence: Selected Prose*. One of America's most respected poets and teachers, she recognizes that "truthfulness, honor, is not something which springs ablaze of itself; it has to be created between people." Like the quality of gift, our honor—being true to ourselves as individuals or as a people—demands healthy exchange to circulate among us.

Perhaps we can trust this in ourselves. While we deal with the trials and strains of our economy and frustrations with our leaders, a reassuring resurgence of solidarity has surfaced like a strong root visibly asserting its growth. Confident again, the strength of our diversity is speaking more easily, with its multiple nuances and vigor in discussions that tune our thinking and aid our mutual acceptance. These are signs of our culture's buoyant root "genius," to use Lewis Hyde's term—a guiding spirit within us that need not perish as long as the spirit of gift is respected.

A living culture! I laugh to myself at the realization again that a living culture, as contentious as it may be, is a rich and unruly beast! Scary but invigorating, challenging but hopeful. Ours carries with it a powerful dream, "one from many." If we are to be a truly inclusive society, as we must be because our twenty-first century is so complex, we cannot expect our differences to melt away. They haven't—they won't—because our visceral identities are expressed through them. Fortunately, in some areas we are giving rein to the dream, learning to shake off our habits of facile attacks and those guilty or romanticized stereotypes that interfere with our listening, and engaging in thinking and caring exchange. Some say this is part of the paradigm shift inside our country. Surely it is the democratic process testing itself again and again. Either way, it is the people—as unafraid, determined, insightful, and respectful as we can be—who will bring about the changes that allow a free society's survival. It is not too late to refurbish a tattered e pluribus unum, our particular human possibility, the fabric of who we are.

Our democracy has lost ground, water, homes and incomes, honesty and spirit. Violence continues to engender violence within our country and beyond its borders. Faced with today's deep insecurities, our imaginations, dialogues, and innovations for a peaceful society are even more necessary. And they are alive. The unifying force of good leadership revealed that, now, much still can be done to re-knit the raveled garments of our communities. The fabric still holds promise. Beneath the halls of the power spectrum, people are seeking, yearning for a common vision that leads us beyond our well-known lowest common denominator of money and its accompanying physical and spiritual impoverishment. In our unabashed yearning, the dream of our forefathers is clearly shaping us again. This time we are maturing. Moving forward, we must abandon not the knowledge of the harm we have done to each other and the earth but the despair that accompanies our awareness.

"Love is . . . the sternest necessity," said poet Peggy Pond Church, who roamed, rode, and lived on the remote Pajarito Plateau north of Santa Fe, deeply imbued with the joy and love of the wild until the US government appropriated her father's Los Alamos Boys' Ranch and built the Manhattan Project, where the atomic bomb was soon invented. "Love is of the sternest necessity." I agree with a friend who states, "This sentence—uttered with such courage despite the grief of her childhood home being lost to a project of mass destruction—can give us all heart." Our capacity for destruction remains great. We are confronted with tough love today. Survival love. The river of our cultural transformation is necessarily deep; our continent is a wide world. Uneroded shores may not be reached in our time, but people, both unknown and famous, are working courageously, sometimes jubilantly, for the crossing, taking up the stern necessity in our belabored land still vulnerable to dissension and greed.

Rested, restored by the mountain and the Taoseños, I get behind the wheel and return to Santa Fe. No longer the grubby nothing of a town Kit Carson had seen, no longer the city struggling with transition I had tried to come home to, it has clearly given itself direction as an urban capital. I'm willing to accept it as it is. Life goes on here with or without me, but happily, I am still able to return for months at a time. And there is good news.

As of January 1, 2009, all for-profit and nonprofit businesses registered with the city are legally obligated to pay their workers a living wage of $9.92 an hour. I am ever so happy for my nieces and nephews and their children that they won't have to scramble quite so hard to eat! The state has brought improvements too. Given that 47 percent of Santa Fe workers live outside the capital due to the high cost of its homes, the Rail Runner, a handsome commuter train, began its run between Albuquerque and Santa Fe the same winter. Adopted with pride and delight by Santa Feans, it will ease the environmental impact of this mobile workforce.

In 2011, a few construction cranes catch your eye as you do errands in town. They look so dissonant in the town's adobe landscape that it's hard to grasp the necessity for them, but it's real. The town is chock full of cars—clean, shiny, newer models. A sign of less rural life and more concentrated wealth, they are in sharp contrast with the blend of funky, dusty cars that meandered through the streets a short ten years ago. Cisco would have a hard time nowadays, though. On the streets and even on the mountain trails where he accompanied me, all the dogs are leashed!

"Old lady nature" is still being squared off for developments in the south end but I know where to find her: either a few minutes above the studios in the Santa Fe Canyon Preserve or outside the house I've just pulled up to, where a friend has offered me a reasonable rent for my room again this year. As I get out of the car, I linger over an opulent yucca in bloom, the straight stalk laden with vanilla and burgundy flowers. I am so happy to be here for this blooming season. I prolong the pleasure. Two perky Gambel's quail scurry purposefully among the abundant chamisa; a rabbit bounds across the road before me.

Is the Santa Fe I loved gone? It is a different place, but I seek out pockets of refuge like this neighborhood where people still like dirt roads, black and starry nights (without bright porch lights), and houses set at a discreet distance from one another with traditional adobe walls or coyote fences for privacy. I don't know all the neighbors, but most of them wave as you go by; some don't, and I'm used to that now. The majority of families have been here for thirty or more years and they are not selling. It's refreshing. No Sotheby's or other real estate signs—so numerous in other neighborhoods—hang in these front yards. The dirt roads are, I admit, bounded a block away by asphalt. But on that asphalted road, as you turn toward the nearby shopping center, a llama farm extends the rural feel.

Dear Grandmother West is dead now; so is Cisco and both my parents, whose generosity to their daughters in dying allows my trips home. Peg's children have had some of their own, and I'm delighted with the newest grandnephew. Among my friends, we did lose Isabelle to her despair in the 1990s. A few others stayed alive and moved elsewhere, but on the whole I'm blessed to be welcomed again and again by a community that has remained intact and grown.

This spring my friend Jim guides me on a walk through the thoroughly renovated railyards. Santa Fe truly has taken a leap into the modern era! The Santa Fe Farmers' Market Pavilion stands proudly at one end. With its shops and a few restaurant bars, the area is enjoyable and stimulating in a very cool, urban style. The new buildings with their squared-off architecture give a hard but clean edge, a mostly tan look, which may be softened by the green when the trees and vines that have been planted grow. There is a park for children to play in. The Acequia Madre, protected, runs through it. Jim is leading me to the discovery of new bike trails—bike trails! the recognizable urban trademark—and a city plan to extend and connect them. The train goes by. We wave but the commuters, not too numerous at 6:30, seem to be occupied with computers and papers in their well-lit cars.

For dinner, we opt for our funky and friendly old Santa Fe atmosphere on the Cowgirl Bar and Grill's patio. Afterward, still on foot, we wander back from there toward Jim's neighborhood, stopping on the shore of the Santa Fe River to watch it rushing, its newly restored banks sprouting fresh willow, the urban atmosphere almost quieted. We chat about rain.

What land would I save?

I'd save Santa Fe, bring the homes back to the people.
But, actually I'm saving the cabin that doesn't exist anymore.
The cabin that means
the silence of the land, the shine of the meadow,
the meanders of the creek woven through the valleys
the stalwart elegance of the columbine
the pure thirst for water
the cold grip of it on rocks and skin in the morning
the spider web that has no porch pole left to weave on
the trails—deer and human—that led us for years through the smell of
pine and over the scrubby dry tracks of the bear

unafraid

The biologist-philosopher Lewis Thomas, in *The Fragile Species*, argues for recognition of a human biological mechanism of pleasure as "a special, independent, autonomous sense. . . . A mechanism that can handle the inside of the messages conveyed both by Shostakovich's more cheerful Fourteenth Quartet and Beethoven's fourth movement of the *Missa Solemnis*, where the violin and the human voice suddenly turn into a single voice, and install the receptor for that word in that line of that poem, that jolt of that image." He encourages his reader to "take into account the need of an organism to know for sure that it is alive." What a daring thought in our work-focused and often violent culture!

Dare we believe that healing the rift of isolation between people and the land, reconnecting with the pleasurable vitality of cultural voice is also essential for our species? Necessary for it to remember that it is alive? Necessary for it to motivate and empower itself to stop the global self-destruction our world is engaged in? Necessary for it to reenter the planet's interconnected web of life? Born to live from "dust to dust" like the deer, the lichen on rock, the dancing crane, we are creatures of this earth. How I hope for our songs to be woven among the whispering trees, nourishing like the rippling rain. Dare we? Dare we sing, dance, chant, tell stories to be believed?

More recently, I've been reading the work of another scientist, Daniel J. Levitin, a leader in music cognition at McGill University. In his book *The World in Six Songs: How the Musical Brain Created Human Nature*, he traces our bonding through music to our own species' (*Homo sapiens*') ancestor, *Homo erectus*. Ages ago, he confirms, "a fifth and crucial use of music was for easing tensions within the larger social groups that were forming—group cohesion." Social groups forming, yes. In today's world, he reminds us, "Humans around the world report not just strong emotional bonding from synchronized, coordinated movement together, but feelings of a spiritual nature—a sense of there being a collective consciousness, the presence of a superior being, or an unseen world that is larger than what we immediately experience."

But, one might ask, what do we have to sing about? Perhaps we can look into the imagination of a scientist friend who reminds us of the deep field of space visible through the Hubble telescope, the actual view of galaxy after galaxy, tens of billions of them moving away from each other in the expanding universe of which our Milky Way galaxy, containing our tiny blue planet, is a fragile and humble part. This is, perhaps, the "Great Sound" the mystic Indian poet Kabir speaks of. We belong too to the beauty of this cosmos.

Coming back to earth, we can sing our fears and sorrows—and our outrage—at the gap between our beauty and wholeness and the ugliness of a violent world: our societies' aggressions on the earth and its people, especially the children who have been bombed; the injustices in our actions and transactions; and always "the egotistical desire," as Wendell Berry fears, that makes us stumble. But being faced with this, we can sing our courage and believe with the people of Mali who repair "wounded objects"—bicycles, cooking pots, ritual objects, or other broken things—that creation, fabrication, and repairing are one and the same human activity. "We repair ourselves every day," says the elder Mali gentleman in a video at the exhibit "Wounded Objects" at the Museum of Civilization in Quebec City. He knows this in the ancestral links in his bones and culture. Humans, we heal. We re-create. We can sing together. We can endure.

On our planet, the necessary chemistry for Lewis's hopeful pleasure "mechanism" is sustained by our awareness of our bonds as men and women, adults and children, human and animal, human and earth. Embracing the knowledge of our belonging to the beauty of the cosmos, we can more easily cherish our world, the remaining wild, the animals, the forests and deserts, and those people with whom we stand, move, and act—be they men, children, women, Indian, Arab, Hispanic, Black, Jew, rich, or poor. To care. We must care.

Aching to be with my people, I had watched the celebration of Obama's election in Chicago on television from across the 49th parallel with a Canadian friend who also wept with relief as we heard his presidency confirmed. We knew the collective elation we were a part of was real, substantive, a deep carrier of new possibility and of futures to live for, to work for. Within my private aching to be at home in the United States came an instant wave of peace through my core. We the people, I sighed to myself, deeply relieved. We the people, I repeated, my heart lilting. We knew at that moment what it meant to be alive again as a people.

In this immense release I couldn't help but remember my dark-haired younger self and her feelings again. Inspired by Lewis Thomas's thoughts on the birth of language, she had written:

> "New, wild, tumbling" sounds can be heard in today's critical mass of concerned women and men, constantly talking, thinking, complaining, wailing, soothing, laughing, regretting, reaching out, bonding, or staying steadily solitary in the face of the continued severe cruelty of the patriarchy and its perversions of power. These sounds are a jubilant, guttural activity below the surface and between the seams of our culture's loud, glossy, media images of life. If the rumblings are not always exultant, it is that isolation and despairing anger still run through them. We must hear the rumblings and the quiet solitary sounds. They come from us. They are our future. We are its hope.

By the night of November 4, 2008, more and more people had heard those sounds. For years we had listened and worried. Then on television, amid the intense jubilation in Grant Park, a reporter asked a man standing in the crowd if Obama would win. Steady, calm, his eyes half closed, almost meditative, shaman-like, the man repeated, "Yes. This is the ground swell, this is the ground swell." I recognized it. The critical mass had grown to a groundswell, reached out through veils of prejudice and beyond fear, bringing this celebration and a new possibility for participative democracy to return to our country. This vibrant, peaceful upheaval had elected a black man as president. My heart remembered my father and was at peace for him, a steady man who never gave up his conviction that in his United States with the spirit of its laws, proper justice could be done.

Like millions of people in the States, Kenya, and around the world, we spoke and sang that night and danced for days in homes, on YouTube, on the phone, and in e-mails. Children, teenagers, students, the very old, and the in-between; men, women, black, Native, Arab, Hispanic, white enjoyed the cohesive hope that had arisen from the groundswell. "*Tous les espoirs sont permis,*" "All our hopes are set free," my Quebecois friends said. Optimism had its force again. We had given ourselves an opportunity to carry forward our ideals and guide us away from greed as our motor. Fellowship reigned. Americans and people abroad alike trusted in a new president capable of giving thoughtful and committed service "for the people." We stood united and jubilant at his inauguration.

A year later, we doubted our president. Politicians wrangling in Congress had thwarted the effectiveness of our groundswell of hope and our once innovative democracy. Will they and corporate leaders eventually choose to help or still hinder the peace and well-being of our society?

New movements fueled by old prejudices and racism are making head-lines. Our frustrations with the climate of dissension make us vulnerable to a renewed fabric of fear that threatens our culture from the inside. We the people are struggling in difficult economic conditions to get beyond the deep resistance to change in ourselves. Will our will hold?

What becomes of our culture's genius does still depend on us and on our efforts to reinvent deliberately, steadily, a human society that under-stands our freedom is measured not only in material wealth, that accepts our children's essential need for nature, and that tends the earth knowing it is our home and all that we have.

De Tocqueville gave weight to the "manners" of the Americans of his time whose moral and intellectual strength he admired. Without the "manners"—the capacity to be innovative while holding moral ground— he felt the country would be lost to those who would take advantage of its laws. This we know has happened through the plethora of lawsuits in our courts and the unwarranted wars we have supported. But the future lies ahead. It is our human relating and interrelating, the sum of our human understanding, our ways and words that create, re-create, or diminish our culture. In re-creating it we need to remind one another that we are alive. Be glad.

And be grateful. Savor our food, each other's touch, the song of the meadowlark, the resilient grasses, moist dust, and gnarly rock of the earth. Rest in the silence of the stars. Repose in knowing there is mystery beyond us. It is my hope that we will not forget our affection for ourselves as simply embodied beings, while we draw on our capacity to shape a society where adults and children alike know they belong, where they are free to experience their exuberance in what it means to be human. Moving forward we may still be fettered, sometimes unwittingly, by the

assumption of our rightness and uniqueness, by our American way's quiet perception that, in spite of our claims to tolerance, all other diverging ways and lives are not quite real, not quite valid. Instead, though, we can, if we will, give ourselves a place to be honestly with our differences, to embrace the naturalness of body and our relatedness to the earth— not theoretically, but in the daily, concrete preciousness of it—then proceed to re/member this world.

In Taos Pueblo, as in other Indian communities, Native American cultures still carry the beliefs of the times when this continent was a homeland for the human spirit. In spite of the torn state many of their people find themselves in, the healthy Native communities offer strength and resilience to their people, nurturing identity based on cooperation and respect for the earth. A spiritual sustenance that consumerism can never offer.

Recently I came across an ad published by the American Indian College Fund. A young Diné woman stands with simple beauty and pride, dignified in the great dry expanse of her people's traditional land, with the Shiprock formation rising behind her. What strikes one is the whole integrity of the woman and her land as she states that by staying on the reservation, she has 80 percent more chance of not dropping out of college. Statistics in the ad also state that tribal colleges foster 90 percent of Indian students completing their studies. The young Navajo's calm beauty adorned with her people's jewelry and her dignified stance in the land itself evoke the traditions of art and community ritual she is part of. She is familiar, strong, determined, at ease, this woman of the West.

Through my Native friends, students, and fellow artists I know this can be. Identity with the earth augurs well for the future.

But on many other reservations, indeed in cities, and even in many rural areas, not everyone has easy access to a community that remains intact in its land. Our media culture leaves most of us bereft of tradition and customs of the kind that the Diné and Pueblo, the Hispanic and Anglo peoples of the Rocky Mountain /Río Grande corridor still have access to. We all want a home that lives up to that name, but due to overdevelopment most of us have little or no access to land vibrant with history and sacred ritual. Urban or rural, wealthy or scrappy survivor, old-timer or newcomer, we are in a common predicament.

And in a global world. The now common iPhone, a scientist friend tells me, holds more computer power than scientists had available to them at the Manhattan Project site, where they invented the atom bomb. Years later, in quiet laboratories, electrical impulses and neuromuscular connections have been proven between people; new awareness of our experience as cellular beings in the universe has been gained; and images of the Big Bang, that first void coalescing into matter, continue to be witnessed on scientists' computer screens.

The world over, as in North America, we go about our daily lives on a planet enlivened, broadened, and impacted by the age of quantum mechanics and space shuttles, even cosmic tourism. Floating in space, like fellow astronauts before him, Quebec's Guy Laliberté, the space tourist, expressed amazed affection for the delicate beauty of the earth turning in the dark, starry void around him and declared the privilege it is to live on planet earth. I had joined friends and thousands of others around the globe for the pleasure of watching *Moving Stars and Earth for Water* on our televisions or computers. Laliberté, also founder of the

Cirque du Soleil, had coordinated these two hours of song, dance, and touching story from the International Space Station to benefit consciousness about our planet's need for clean water. The last segment, a short conversation between the astronaut Julie Payette on earth and the astronauts aboard the ISS, gave me my first sense of how real the human experiences and experiments already created and shared in this controlled environment beyond the earth are. For over two decades after the first flight in a Montgolfière in 1783, the scientists and poets alike who flew in the balloons wrote extensively about the moving beauty of the earth, its forests and oceans and the contiguous land they had discovered from the skies. Perhaps we are reaching a time again when the wonders of our planet and our love for it can be mutually supported by science and art.

Along with the blue planet easily visible now on our own computer screens, I marvel that beliefs similar to the ancient Hindu *prana*—"vital air" or "the life-sustaining force"—and Native peoples' sense of breath and wind as our animating force are being validated by scientific consciousness and research. I'm reassured that more recent scientific thought on the emergent universe and our belonging to it dovetails with such beliefs as well as the *sympatico* understood among humans for centuries. Both ancient myths and scientific awareness inform our human lives. A generic root, DNA, confirms our relatedness. Physics, mythology, tradition, and creativity studies intersect in the insights they are offering us. The new complexities of our world offer a time of hope, new realization, and discovery. Yet, paradoxically and frighteningly, it remains a time of fragile expectation for life on the planet. Our wealth has not protected the lives of our soldiers at war. Or our children at home. More than ever, our culture needs well-informed habits of the heart.

Selected Bibliography

INVITATION TO AN EXPANDING LITERATURE

Abley, Mark. Spoken Here: *Travels among Threatened Languages*. New York: Houghton Mifflin, 2003.

Atencio, Ernest. *La Vida Floresta: Ecology, Justice and Community-Based Forestry in Northern New Mexico*. Santa Fe, NM: Northern New Mexico Group of the Sierra Club, 2004.

_____. *Of Land and Culture: Environmental Justice and Public Lands. Ranching in Northern New Mexico*. Santa Fe, NM: The Quivira Coalition and the Northern New Mexico Group of the Sierra Club, 2004.

Berry, Thomas. *Evening Thoughts: Reflecting on Earth as Sacred Community*. Berkeley, CA: Sierra Club Books and University of California Press, 2006.

Berry, Wendell. *A Continuous Harmony, Essays Cultural and Agricultural*. San Diego: Harcourt Brace Jovanovich, 1970.

_____. *Given*. Berkeley, CA: Counterpoint, 2006.

_____. *The Mad Farmers Poems*. Berkeley, CA: Counterpoint, 2008.

Black Elk. *Black Elk Speaks: Being the Life Story of a Holy Man of the Oglala Sioux, as told through John G. Neihardt*. New York: Washington Square Press and Pocket Books, 1972.

Brown, Dee. *Bury My Heart at Wounded Knee: An Indian History of the American West*. New York: Bantam Books, 1972.

Butala, Sharon. *The Perfection of the Morning: An Apprenticeship in Nature*, Toronto: Harper, 1994.

Carson, Rachel. *Silent Spring*. Boston: The Riverside Press Cambridge, Houghton Mifflin Company, 1962.

Church, Peggy Pond. *Church: New and Selected Poems*. Boise, ID: Ahsahta Press, !976.

deBuys, William. *The Walk*. San Antonio, TX: Trinity University Press, 2007.

deBuys, William, and Don J. Usner. *Valles Caldera: A Vision for New Mexico's National Preserve*. Santa Fe, NM: Museum of New Mexico Press, 2006.

de Tocqueville, Alexis. *Democracy in America*. Translated by Henry Reeve. New York: Alfred A. Knopf, 1994.

Didion, Joan. *Where I Was From*. New York: Random House, 2003.

Dippie, Brian W. *"Paper Talk," Charlie Russell's American West*. New York: Alfred A. Knopf, 1979.

Forrest, Suzanne. *The Preservation of the Village: New Mexico's Hispanics and the New Deal*. Santa Fe, NM: University of New Mexico Press, 1989.

Griffiths, Jay. *Wild: An Elemental Journey*. New York: Jeremy P. Tarcher, Penguin, 2006.

Hallendy, Norman. *Inuksuit. Silent Messengers of the Artic*. Vancouver: Douglas McIntyre Ltd., 2001.

Hundertwasser, Friedensreich. *Hundertwasser*. Miniature exhibit catalogue, 3rd ed. Glarus/Switzerland: Greuner Janura, 1973.

Hyde, Lewis. *The Gift: Imagination and the Erotic Life of Property*. New York and Toronto: First Vintage Books Edition, 1983.

Josephy, Alvin M. Jr. *500 Nations: An Illustrated History of North American Indians*. New York: Alfred A. Knopf, 1994.

Kittredge, William. *Hole in the Sky: A Memoir*. New York: Alfred A. Knopf, 1992.

————. *The Nature of Generosity*. New York: Random House, 2001.

_____. *Owning It All: Essays.* Saint Paul, MN. Graywolf Press, 1987.

Leopold, Aldo. *A Sand County Almanac, with Essays on Conservation from Round River.* New York: Ballantine Books, 1966.

Lethem, Jonathan. "The Ecstasy of Influence: A Plagiarism." *Harper's* (February 2007): .

Levitin, Daniel J. *The World in Six Songs: How the Musical Brain Created Nature.* Toronto: Penguin, 2008.

Limerick, Patricia Nelson. "The Rise of Regret in the American West: How to Tell Meaningful Action from Pointless Wallowing." Lecture presented at the University of Colorado, 2003.

Lopez, Barry. *Arctic Dreams: Imagination and Desire in a Northern Landscape.* New York: Bantam Books, 1987.

Louv, Richard. *Last Child in the Woods: Saving Our Children from Nature-Deficit Disorder.* Chapel Hill, NC: Algonquin, 2006.

Manion, Patricia Jean, SL. *Beyond the Adobe Wall: The Sisters of Loretto in New Mexico.* Independence, MO: Two Trails Pub Press, 2001.

McFadden, Steven. *The Call of the Land: An Agrarian Primer for the 21st Century.* Bedford, IN: Norlights Press, 2009.

Merwin, W. S. *The Rain in the Trees.* New York. Alfred A Knopf, 1989.

Momaday, N. Scott. *House Made of Dawn.* New York: Harper and Row, 1968.

_____. *The Names: A Memoir.* Tucson, AZ: Suntracks/The University of Arizona Press, 1976.

_____. "Sacred Places." Calendar frontispiece. San Francisco: Sierra Club Publications, 1994.

Nash, Kate. "King: Political Philosophy Grew Out of Upbringing on Rural Homestead." *The Santa Fe New Mexican*, November 14, 2009.

Notes on the Acequia Madre de Santa Fe. Santa Fe, NM: Community Acequia Association, 2007.

Oakes Baile, ed. *Sculpting with the Environment: a Natural Dialogue.* New York: Van Nostrand Reinhold,1995.

Pierson, Melissa Holbrook. *The Place You Love Is Gone: Progress Hits Home.* New York: W. W. Norton. 2006

Pritchett, Laura, Richard L. Knight, and Jeff Lee. *Homeland: Ranching and a West That Works.* Boulder, CO: Johnson Books, 2007.

Rich, Adrienne. *On Lies, Secrets and Silence.* Selected Prose 1966-1978. New York: W. W. Norton & Company. 1979.

Sayre, Nathan F. *Working Wilderness: The Malpai Borderlands Group and the Future of the Western Range.* Tucson, AZ: Rio Nuevo Publishers, 2005.

Sharpe, Tom. "Birth of Historic Style." *The Santa Fe New Mexican*, October 23, 2006.

Simmons, Marc. *New Mexico: A History.* New York: W.W. Norton and American Association for State and Local History, 1977.

Simon, Sherry. *Translating Montreal: Episodes in the Life of a Divided City.* Montreal: McGill-Queens University Press, 2006.

Snyder, Gary. *The Practice of the Wild: Essays.* New York: North Point Press, 1990.

Sojourner, Mary. *Bonelight: Ruin and Grace in the New Southwest.* Reno, NV: University of Nevada Press, 2002.

Sullivan, Julie. "An Ecotone, Not a Divide." *The Quivira Coalition Journal*, October 2006.

Suzuki, David, and Amanda McConnell. *The Sacred Balance: Rediscovering Our Place in Nature.* Vancouver: Douglas and McIntyre, 2002.

Travis, William R., D. M. Theobald, G. W. Mixon, and T. W. Dickinson.

Western Futures: A Look into the Patterns of Land Use and Future Development in the American West. Report from Center of the American West, 2005.

Thomas, Lewis. *The Fragile Species.* New York: Charles Scribner's Sons, 1992.

Turner, Frederick. *Beyond Geography: The Western Spirit Against the Wilderness.* New Brunswick, NJ: Rutgers University Press, 1990.

Vaillant, John. *The Golden Spruce: A True Story of Myth, Madness and Greed.* Toronto: Vintage Canada, 2005.

White, Courtney. *The Quivira Coalition: Fresh Eyes on the Land, Innovation and the Next Generation. Conference Program.* Santa Fe, NM: The Quivira Coalition, January 2007.

_____. *Revolution on the Range: The Rise of a New Ranch in the American West.* Washington DC: Island Press, 2008.

Whitlock, Flint. *Turbulence Before Takeoff: The Life and Times of Aviation Pioneer Marlon Dewitt Green.* Brule, WI: Cable Publishing, 2009.

Whitman, Walt. *Edward Weston, Leaves of Grass.* Reprint of the 1942 edition published by the Limited Editions Club. Binghampton, NY: Vail-Ballou Press, 1970.

Williams, Terry Tempest. *Finding Beauty in a Broken World.* New York: Pantheon, 2008.

_____. *Red: Passion and Patience in the Desert.* New York: Pantheon, 2001.

_____. *Refuge: An Unnatural History of Family and Place.* New York: Vintage Books, 1991.

Internet Resources

www.acoustic, Ecology.org/wildlandbiology, "The Biological Effects of Noise on Wildlife" by Bernie Krause

www.ameinfo.com, Middle East business resource

www.americantowns.com

www.archives.cbc.ca

www.cbc.ca/radio

www.centerwest.org, Center of the American West

www.centerforfoodsafety.org

www.edd.state.nm.us Census Profiles for New Mexico

http://quickfacts.census.gov

www.newworldencynclopedia.org, Some statistics for demographics

www.encartamsnencyclopedia

www.nmcn.org/heritage/civil_war/essays, Charles Bennett, The Civil War in New Mexico

www.nps.gov

www.malpaiborderlandsgroup.org

www.ogj.com, Oil and Gas Journal

www.pickensplan.com, T. Boone Picken's wind development

www.quiviracoalition.org

www.santefenewmexican.com, *The Santa Fe New Mexican* (newspaper) www.thesunmagazine.org/issues/417/quiet_please, "Quiet, Please." An interview with Gordon Hempton on acoustic ecology

www.theodoreroosevelt.org

www.utecreekcattlecompany.com, Ute Creek Cattle Company, Tuda Libby Crews and Jack Crews

www.verticalfarming.com, Dr. Dickson Despommier on Vertical Farming

www.westernfolklife.org, Elko Poetry Festival

www.wildsanctuary.com, Bernie Krause and bio-acoustics

About the Author

Born and raised in Colorado and introduced at age ten to New Mexico, Rae Marie Taylor early on cultivated an appreciation for the beauty and cultural diversity of the American Southwest, as well as a sense of shared habitat and kinship with its vegetation and wildlife. After finishing college in Denver, her passion for the French language drew her to Quebec, Canada, where she later founded the first courses on Native American literature at Montreal's Dawson College and Concordia University's Simone de Beauvoir Institute. While teaching in the province of Quebec, she also became an artist and poet.

Keenly interested in petroglyphs and her own roots, Rae Marie journeyed back to the Southwest and was soon invited to be an illustrator for Mesa Verde National Park's archaeology lab. Staying in the region for several years, she engaged in fieldwork for the former Santa Fe Regional National Park Office and eventually served as conference coordinator and guide in Southwest art and archaeology for Recursos de Santa Fe.

Following a return to teaching in Quebec, she produced the Spoken Word CD *Black Grace* with Montreal musician David Gossage. Whether recording, performing in festivals and other poetry venues, or starring in one-woman shows—her *Chant du Nord, regard du Sud* at L'Espace Félix-Leclerc in Quebec and *An Earthly Hour: A Human Time at the Loretto Chapel* in Santa Fe, New Mexico— her work has consistently celebrated the cultural and natural beauty of her Southwestern homeland.

Most recently, she was invited to perform in the *Poetry Jam* at the Lensic Theater, at the Santa Fe Literary Center, and in other popular

Northern New Mexico venues. In addition, her essay "Release" was published in the 2011 anthology *The Return of the River: Writers, Scholars, and Citizens Speak on Behalf of the Santa Fe River.*

Rae Marie's migrations north and south, between Quebec's wooded and well-watered landscape and the high, dry Rocky Mountain/Río Grande corridor, have broadened her perspective on the complex interactions of habitat and culture. Vitally concerned about the impact of development on land and water, she bears witness in her book of essays, *The Land: Our Gift and Wild Hope,* to both their devastation and today's resurgent hope for renewal.